The Haynes Chevrolet Engine Overhaul Manual

by Robert Maddox and John H Haynes
Member of the Guild of Motoring Writers

The Haynes Automotive Repair Manual for overhauling Chevrolet V8 engines

(10305-7X12)

Haynes Group Limited
Haynes North America, Inc.
www.haynes.com

Acknowledgements

We are grateful for the help and cooperation of Performance Automotive Wholesale (PAW) in Chatsworth, California who graciously supplied the parts needed to overhaul our two project engines. The Super Shops in Thousand Oaks, California allowed us to photograph the performance parts shown in Chapter 9. Machine shop work was also done by PAW.

© **Haynes North America, Inc. 1991, 1999**

With permission from Haynes Group Limited

A book in the Haynes Automotive Repair Manual Series

ISBN-10: 1-85010-762-9

ISBN-13: 978-1-85010-762-0

Library of Congress Catalog Card Number 91-71154

Contents

Chapter 6 Overhauling the engine block

Chapter 7 Reassembling and installing the engine

Chapter 8 Related repairs

Chapter 9 Improving performance and economy

Glossary

Appendix

Index

 **Introduction**

How to use
this repair manual

The manual is divided into Chapters. Each Chapter is sub-divided into Sections, some of which consist of consecutively numbered Paragraphs (usually referred to as "Steps", since they're normally part of a procedure). If the material is basically informative in nature, rather than a step-by-step procedure, the Paragraphs aren't numbered.

The first three Chapters contain material on preparing for an overhaul. The remaining Chapters cover the specifics of the overhaul procedure.

Comprehensive Chapters covering tool selection and usage, safety and general shop practices have been included.

The term "**see illustration**" (in parentheses), is used in the text to indicate that a photo or drawing has been included to make the information easier to understand (the old cliche "a picture is worth a thousand words" is especially true when it comes to how-to procedures). Also, every attempt is made to position illustrations directly opposite the corresponding text to minimize confusion. The two types of illustrations used (photographs and line drawings) are referenced by a number preceding the caption. Illustration numbers denote Chapter and numerical sequence within the Chapter (i.e., 3.4 means Chapter 3, illustration number four in order).

The terms "**Note**", "**Caution**", and "**Warning**" are used throughout the text with a specific purpose in mind - to attract the reader's attention. A "**Note**" simply provides information required to properly complete a procedure or information which will make the procedure easier to understand. A "**Caution**" outlines a special procedure or special steps which must be taken when completing the procedure where the **Caution** is found. Failure to pay attention to a **Caution** can result in damage to the component being repaired or the tools being used. A "**Warning**" is included where personal injury can result if the instructions aren't followed exactly as described.

Even though extreme care has been taken during the preparation of this manual, neither the publisher nor the author can accept responsibility for any errors in, or omissions from, the information given.

What is an overhaul?

An engine overhaul involves restoring the internal parts to the specifications of a new engine. During an overhaul, the piston rings are replaced and the cylinder walls are reconditioned (rebored and/or honed). If a rebore is done, new pistons are required. The main bearings, connecting rod bearings and camshaft bearings are generally replaced with new ones and, if necessary, the crankshaft may be reground to restore the journals.

Generally, the valves are serviced as well, since they're usually in less-than-perfect condition at this point. While the engine is being overhauled, other components, such as the distributor, starter and alternator, can be rebuilt as well. The end result should be like a new engine that will give many thousands of trouble-free miles. **Note:** *Critical cooling system components such as the hoses, drivebelts, thermostat*

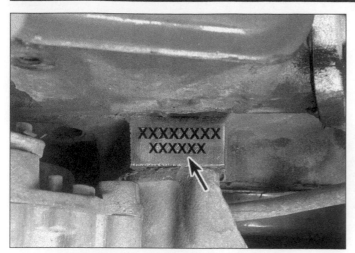

1.1 Typical Chevrolet code number locations

1.2 Check for a casting number at the rear of the block (arrow)

and water pump MUST be replaced with new parts when an engine is overhauled. The radiator should be checked carefully to ensure that it isn't clogged or leaking; if in doubt, replace it with a new one. Also, always install a new oil pump when overhauling the engine - we recommend against rebuilt pumps.

It's not always easy to determine when, or if, an engine should be completely overhauled, as a number of factors must be considered.

High mileage is not necessarily an indication that an overhaul is needed, while low mileage doesn't preclude the need for an overhaul. Frequency of servicing is probably the most important consideration. An engine that's had regular and frequent oil and filter changes, as well as other required maintenance, will most likely give many thousands of miles of reliable service. Conversely, a neglected engine may require an overhaul very early in its life.

Excessive oil consumption is an indication that piston rings, valve seals and/or valve guides are in need of attention. Make sure that oil leaks aren't responsible before deciding that the rings and/or guides are bad. Perform a compression check (see Chapter 3) and have a leak-down test performed by an experienced tune-up mechanic to determine the extent of the work required.

If the engine is making obvious knocking or rumbling noises, the connecting rod and/or main bearings may be at fault. Check the oil pressure with a gauge installed in place of the oil pressure sending unit or switch (see Chapter 3) and compare it to the specifications for the particular engine (see the *Haynes Automotive Repair Manual* for your vehicle). If it's extremely low (generally, less than 10 psi at idle), the bearings and/or oil pump are probably worn out.

Loss of power, rough running, excessive valve train noise and high fuel consumption rates may also point to the need for an overhaul, especially if they're all present at the same time. If a complete tune-up doesn't remedy the situation, major mechanical work is the only solution.

Before beginning the engine overhaul, read through this entire manual to familiarize yourself with the scope and requirements of the job. Overhauling an engine isn't partic-

ularly difficult if you have the correct equipment; however, it is time consuming. Plan on the vehicle being tied up for a minimum of two weeks, especially if parts must be taken to an automotive machine shop for repair or reconditioning. Check on availability of parts and make sure that any necessary special tools and equipment are obtained in advance. Most work can be done with typical hand tools, although a number of precision measuring tools are required for inspecting parts to determine if they must be replaced. Often an automotive machine shop will handle the inspection of parts and offer advice concerning reconditioning and replacement.

Engine identification

Identifying just which engine you have is critical because Chevrolet engines which are very similar in appearance can be quite different in important details.

Prior to 1972 on passenger cars and 1973 on light trucks, the Vehicle Identification Number (VIN) only indicates whether the vehicle originally came with a six-cylinder or a V8. Some of these engines can be identified by the decals on the air cleaner or the valve covers. If these markings are missing or you suspect they are incorrect, check for numbers on the engine. Record the VIN number **(see illustration)** and also the casting number on the rear of the engine block between the distributor and bellhousing **(see illustration)**. Also note the shape of the casting marks on the ends of the cylinder heads (later models don't have these). Using this information, check with your local dealer parts department or salvage yard for assistance in identification.

On 1972 and newer passenger cars and 1973 and newer light trucks, the first step of engine identification is to look at the VIN, because the VIN includes a code letter that indicates which engine the vehicle is equipped with. The VIN plate is located at the left front corner, just inside the windshield on passenger cars or on the door jamb of trucks and vans **(see illustrations)**.

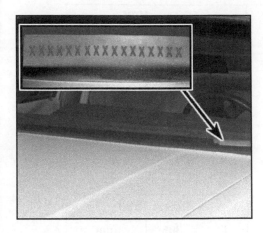

CHEVROLET Passenger Cars

Model Year Codes

2 = 1972	B = 1981	
3 = 1973	C = 1982	
4 = 1974	D = 1983	
5 = 1975	E = 1984	
6 = 1976	F = 1985	
7 = 1977	G = 1986	
8 = 1978	H = 1987	
9 = 1979	J = 1988	
0 = 1980	K = 1989	
	L = 1990	

Example (1972 through 1980 models)

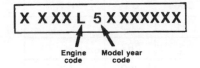

X X XX L 5 X XXXXXX

Engine code — Model year code

Example (1981 through 1990 models):

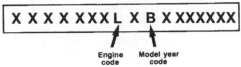

X X X X XXX L X B X XXXXXX

Engine code — Model year code

1972

Code	C.I.	H.P.
F	307	(130 HP)
H	350	(165 HP)
J	350	(175 HP)
K	350	(200 HP)
L	350	(255 HP)
R	400	(170 HP)
S	402	(210 HP)
U	402	(240 HP)
V	454	(230 HP)
W	454	(270 HP)

1973

Code	C.I.	H.P.
F	307	(115 HP)
H	350	(145 HP)
J	350	(190 HP)
K	350	(175 HP)
R	400	(150 HP)
T	350	(245 HP)
X	454	(215 HP)
Y	454	(245 HP)
Z	454	(275 HP)

1974

Code	C.I.	H.P.
H	350	(145 HP)
J	350	(195 HP)
K	350	(185 HP)
L	350	(160 HP)
R	400	(150 HP)
T	350	(245 HP)
U	400	(180 HP)
Y	454	(235 HP)
Z	454	(270 HP)

1975

Code	C.I.	H.P.
G	262	
H	350	(145 HP)
J	350	(165 HP)
L	350	(155 HP)
T	350	(205 HP)
U	400	
Y	454	

1976

Code	C.I.	H.P.
G	262	
L	350	(165 HP)
Q	305	

1976 (con't)

Code	C.I.	H.P.
S	454	
U	400	
V	350	(145 HP)
X	350	(210 HP)

1977

L	350	(170 HP)
U	305	
X	350	(210 HP)

1978

H	350	(220 HP)
L	350	(170 HP)

1979

G	305	(2 barrel)
H	305	(4 barrel)
J	267	
L	350	
4	350	
8	350	

1.3 Passenger car V8 engine codes

1980		
Code	C.I.	H.P.
H	305	
J	267	
L	350	
6	350	
8	350	

1981		
H	305	
J	267	
L	350	
6	350	

1982		
H	305	
J	267	
7	305	(fuel injection)
8	350	(fuel injection)

1983		
Code	C.I.	H.P.
H	305	
S	305	
6	350	(police)
7	305	
8	350	

1984		
G	305	
H	305	
6	350	
8	350	(fuel injection)

1985 – 1986		
F	305	(fuel injection)
G	305	
H	305	
6	350	(police)
8	350	(fuel injection)

1987 – 1988		
Code	C.I.	H.P.
E	305	(fuel injection)
F	305	(TPI)
G	305	(4 barrel)
H	305	(4 barrel)
Y	305	(4 barrel)
8	350	(TPI)

1989 – 1990		
E	305	(fuel injection)
F8	305	(fuel injection)
Y	305	
8	350	(fuel injection)

1.3 Passenger car V8 engine codes (continued)

CHEVROLET & GMC Light trucks

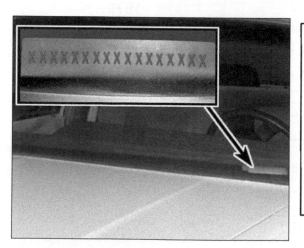

Model Year Codes

4 = 1974	B = 1981
5 = 1975	C = 1982
6 = 1976	D = 1983
7 = 1977	E = 1984
8 = 1978	F = 1985
9 = 1979	G = 1986
0 = 1980	H = 1987
	J = 1988
	K = 1989
	L = 1990

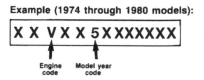

Example (1974 through 1980 models):

X X V X X 5 X XXXXXX

Engine code | Model year code

Example (1981 through 1990 models):

X X X X XXX L X F X XXXXXX

Engine code | Model year code

1.4 Light truck V8 engine codes

1973		
Code	C.I.	H.P.
X	307	(130 HP)
Y	350	(155 HP)
Z	454	(240 HP)
1974		
L	454	(245 HP)
U	350	(export)
V	350	(2 barrel)
W	350	(LP gas)
Y	350	(160 HP)
Z	454	(230 HP)
1975		
L	454	(245 HP)
M	400	
U	350	(export)
V	350	(2 barrel)
Y	350	(4 barrel)
Z	454	(230 HP)
1976		
L	350	
S	454	
U	400	
V	350	
Y	454	
1977		
L	350	
R	400	
S	454	
U	305	
Y	454	

1978		
Code	C.I.	H.P.
L	350	
R	400	
S	454	
U	305	
Y	454	
1979		
L	350	
M	350	
R	400	
S	454	
U	305	
1980		
G	305	
L	350	
M	350	
P	350	
R	400	
S	454	
W	454	
X	400	
1981		
F	305	
G	305	
H	305	
J	267	
L	350	
M	350	
P	350	
W	454	

1982-1984		
Code	C.I.	H.P.
F	305	
H	305	
L	350	
M	350	
P	350	
W	454	
1985-1986		
F	305	
H	305	
K	350	
L	350	
M	350	
N	262	
W	454	
1987		
H	305	(TBI)
K	350	(TBI)
M	350	(4 barrel)
N	454	(TBI)
W	454	(4 barrel)
1988		
H	305	(fuel injection)
K	350	(fuel injection)
N	454	(fuel injection)
1989- 1990		
H	305	(fuel injection)
K	350	(fuel injection)
M	350	
N	454	(fuel injection)
W	454	

1.4 Light truck V8 engine codes (continued)

Parts interchangeability

A lot of time and money can be saved if you know which parts are interchangeable between your engine and those available on the used market and in wrecking yards.

There is considerable interchangeability within the Chevrolet small block and big block families. However, few parts from one family are usable in the other; about the only notable exception is the distributor.

Due to the vast number of Chevrolet V8 engines produced since 1955 in many versions, a complete and comprehensive guide would require several volumes the size of this book. Most wrecking yards have interchange manuals that provide a wealth of information for parts swapping. The following information provides a basic overview, and does not cover every possible combination of parts.

We will treat the small block and big block engines separately; refer to the appropriate Sections based on the engine family you are working on˙

Small block V8s

Small block Chevrolet engines were produced in the largest numbers of any automotive powerplant in history. Beginning with 1955 models through the present, they have been built in 262, 265, 267, 283, 302, 305, 307, 327, 350 and 400 cubic inch versions.

The different displacements are obtained by combining several bores and strokes in various combinations:

Bore (inches)	Stroke (inches)	Displacement (cubic inches)
3.671	3.10	262
3.750	3.00	265
3.500	3.48	267
3.875	3.00	283
4.001	3.00	302
3.736	3.48	305
3.876	3.25	307
4.001	3.25	327
4.001	3.48	350
4.126	3.75	400

Displacement table

Camshafts

Camshaft specifications vary considerably among different model years and horsepower versions. The camshaft specifications should closely match the rest of the engine to assure the best performance, driveability, economy and lowest emissions.

Except for 1987 and newer models with roller lifters, camshafts are interchangeable from one year or model to the other, since the journals have the same diameter and spacing. **Note:** *On 1955 and some 1956 265 engines, the rear journal of the camshaft must be notched to assure an oil supply to the valve gear.*

Chevrolet factory camshafts are available for hydraulic, solid and even roller lifters (on 1987 and later models). Factory roller cams and lifters can only be installed in 1987 and newer blocks that were designed for them. Camshafts and non-roller lifters should always be replaced together as a set.

It is often difficult to determine which camshaft is installed in a particular used engine before it is taken apart. Even after the camshaft is removed, casting numbers are unreliable for identifying specific grinds. Although used camshafts can be installed in another engine if the lifters are reinstalled on the same lobes, we recommend new camshafts and lifters when replacement is necessary.

Cylinder heads

Except for a few aluminum heads found on racing Corvettes around 1961, virtually all factory made small block heads are made from cast iron.

Cylinder heads are available in a wide variety of valve sizes, combustion chamber volume, spark plug type and port size.

Pre-1959 heads can be identified by the spacing of the valve cover bolts. On 1959 and later heads, the bolt holes are directly opposite each other **(see illustration)**; on early

1.5 Most 1959 and later cylinder heads have valve cover bolts that are directly opposite each other

1.6 1969 and later heads have bolt holes (arrows) for accessories

models the spacing was staggered. The latest heads have valve covers with four bolts running lengthwise down the center. Valve covers must be selected to match the type of cylinder head.

All 1955 through 1968 heads lack mounting bolt holes on the ends for accessories such as power steering pumps or alternators **(see illustration)**. Later heads can be used on early models, but not vice versa.

Early in 1971 Chevy changed from straight to slant spark plug locations. The plug design was also changed from flat seat and washer type to tapered seat type. These heads have larger combustion chambers for lower compression.

Several valve sizes were used over the years. Early 265 and 283 heads used 1.72 inch diameter intake and 1.50 inch exhaust valves. Later high performance 283 heads have 1.94 inch intake valves.

For high performance use, the "202" or "fuelie" heads are very popular. These have large ports and 2.02 inch diameter intake valves and 1.60 inch exhausts. They were used mostly on Z28s and Corvettes with 302, 327 and 350 fuel injected and high performance carbureted engines. The present part number for these is 3987376, superseding 3853608, 3958604 and 3928445.

Later angle-plug high performance heads go by part number 3965742. A 1976 head with large spring seats is number 336746. Pre-1974 heads are not designed for unleaded or low-lead fuel and should have special valves and hardened valve seats installed to reduce valve recession on today's fuel.

Production aluminum heads were introduced on 1986 Corvettes. In 1988 they were updated with larger "D" shaped exhaust ports. If these heads are used, be sure to match the intake and exhaust manifolds to the ports.

Due to the low octane fuels currently available, the compression ratio should be limited to about 9 or 9.5 to 1. Many high performance heads have small combustion chamber volumes and high dome pistons that raise the compression to 10.5 or higher. Try to find heads from an engine that did not have extremely high compression and use flat top pistons.

Heads are available in both standard small port and high performance large port versions. Using an intake manifold gasket, compare the size of the ports in the heads and intake manifold to ensure a match. Compare the exhaust manifold mounting bolt pattern to the exhaust manifold/header you intend to use; there are different versions. Always check carefully for valve sizes, combustion chamber volume, exhaust manifold and spark plug type and port size when comparing cylinder heads to obtain the correct ones. Look for cracks and deep-seated valves and reject any worn out or damaged heads. Whenever possible, obtain replacement cylinder heads in matched pairs to ensure they are the same.

Intake manifolds

There are a great many choices in intake manifolds. The factory has offered two-barrel and four-barrel carburetors, dual four-barrel carburetors, Rochester fuel injection (from 1957 through 1965), throttle body fuel injection and port injection.

If you are working on a fuel injected model or restoring a vehicle to original condition, an identical part must be obtained. We will not go into detail on fuel injected or multi-carb models, since the choices are limited. However, for most applications using a single carburetor, several types of manifolds will work satisfactorily.

Although the bolt pattern is the same on all years, early models have an oil filler tube; later models have filler caps in the valve covers. Be sure to use the correct version.

For low-performance work, a cast-iron manifold with a two-barrel carburetor will suffice. Early models generally have smaller carbs and ports than later models. Also, the automatic chokes can be activated by exhaust manifold tubes or choke stoves mounted next to the carburetor.

A single four-barrel carburetor manifold works best for most applications. These are available with mounting provisions for Carter, Rochester and Holley carburetors. Early

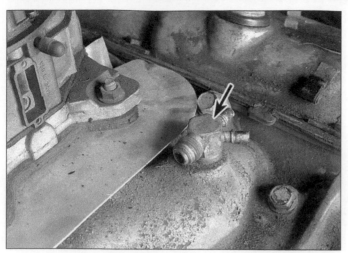

1.7 If you're swapping manifolds, make sure there are provisions for all your vacuum fittings (arrow)

1.8 Crankshafts with a two-piece rear main oil seal look like this from the rear

models and/or lower horsepower engines can use manifold part number 3888886 with a small Carter or Holley four-barrel. From 1966 on, most four-barrel models used Rochester Quadrajets on a manifold with part number 3987361.

Relatively few aluminum four-barrel manifolds were installed at the factory. Exchanging a cast-iron manifold for an aluminum one is an easy way to shave weight off the front end of a vehicle.

For high-performance street use, the factory aluminum high-rise is hard to beat. It mounts a square-flange Holley four-barrel or Carter AFB and goes under part number 3958627 superseded to 3972114.

Always compare the port sizes of the heads and manifolds by holding a gasket up to both openings. For optimum performance, the sizes must match. Larger ports favor high rpm and larger displacements. Small ports tend to improve low rpm torque and driveability.

On 1973 and later models, look for Exhaust Gas Recirculation (EGR) valve provisions. Always compare the old and new manifolds to determine if sufficient vacuum fittings are present **(see illustration)**.

Exhaust manifolds

The choice of exhaust manifolds is limited to a few basic designs; none of these are exceptional from a performance standpoint. Many enthusiasts convert to tubular headers if they are looking for more performance.

Generally, the "ram's horn" shape manifolds used on early Corvettes and high-performance passenger cars produce more power than the log shape designs. However, clearance between body and frame determines which exhaust manifolds can be used in a given situation.

There are several types of exhaust outlet flanges from 2 to 2 1/2 inches in diameter, using both two and three studs. Heat risers or EFE valves, if used, must fit the manifold correctly. Also, accessory mounts and the length and angle of exhaust outlet must match the engine and vehicle. Be sure to check the number of cylinder head-to-manifold mounting bolt holes; these vary on different models.

Crankshafts

There are a vast number of different versions of small block Chevy crankshafts. Over the years, several different stroke lengths and journal diameters have been introduced.

Unique among them are the 400 c.i. models; these stand alone and are not interchangeable with any others. Also, 400 crankshafts are externally balanced and must be used with flywheels/driveplates and harmonic balancers designed for the 400.

Journal diameter remained the same on 265, 283, 327 and early 350s through 1967. Stroke was the same on 265 and 283 models, but 327s have a longer stroke. Beginning in 1968, the journal diameters were increased and these cannot be swapped with earlier models.

High-performance models used forged crankshafts with Tuftrided journals (a special hardening treatment) and lower performance models have cast cranks. Forged crankshafts can be used in low performance models, but cast cranks should not be used in high-revving performance engines. Generally, the newest or highest performance version with a given stroke will be the best choice.

Beginning in 1986, the rear main oil seal design was changed from a two-piece to a one-piece seal. The flywheel mounting flange on the crankshaft was also modified to be compatible with this change **(see illustration)**. Be sure to use all the components from one type together.

Connecting rods

There are three basic sizes of small block V8 connecting rods. These are found in 1955 through 1967, 1968-on (except 400 c.i.) and 400 cubic inch models.

All small blocks, with the exception of the 400 cubic inch version, use the same length connecting rods. The rods in 400 c.i. models are 5.565 inches center-to-center while all other rods are 5.700 inches center-to-center.

However, models prior to 1968 have smaller diameter crankshaft journals and therefore are not interchangeable with later models.

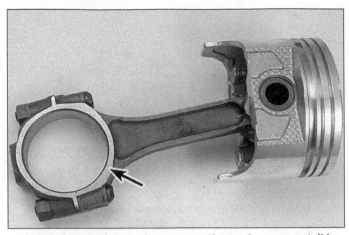

1.9 This is not a heavy-duty connecting rod - you can tell by comparing it to a heavy-duty one: it isn't as thick around the big end (arrow), also, the rod bolts are smaller

High performance engines use heavy-duty connecting rods with larger diameter bolts. These are preferred for high rpm use. Compare the thickness around the big end and also the bolt diameter **(see illustration)**.

Connecting rods should always be installed in matched sets of the same type. Engine balance is affected by rod weight and should be checked by an automotive machine shop any time rods are replaced.

Blocks

There are quite a few different engine block designs in the small block family. Except for a few racing-designed engines, all of the stock blocks are made of cast iron.

Early 265 models were cast without provision for an oil filter. These blocks should only be used when an authentic restoration is desired.

Starting in 1958, all V8s have bosses cast into the side of the block for engine mounts. Some engines also have cast bosses above the oil filter area for clutch linkage mounting. These must be present on replacement blocks - do not attempt to drill into a casting if it lacks the mounting boss.

In 1959 the rear crankshaft oil seal was changed from a rope-type to a neoprene-type seal. Engine blocks remained unchanged until 1962. For 1962, the 327 c.i. was introduced with a larger bore and longer stroke.

During 1967 a 302 was made by combining a three-inch stroke crankshaft with a four inch bore and a 350 was created by stroking a 327 to 3.48 inches. For 1968, the 307 was introduced by using the same stroke as a 327 and same bore as a 283. Also, the diameter of the crankshaft journals was increased on small blocks. Only large journal crankshafts can be installed in later model blocks. Conversely, only small diameter cranks will fit in early blocks.

One of the most highly prized blocks is the 350 with four bolt main bearing caps. This combination was found mostly in high performance models with four-barrel carburetors. These engines provide the strongest bottom end with a fairly large cubic inch displacement.

Beginning in 1970, a 400 c.i. version of the ubiquitous small block was introduced. This engine shares basic design features with other small blocks, but the crankshaft, rods, block and heads are unique to this model.

Later, two low-performance versions were produced with displacements of 262 and 267 cubic inches. These engines were designed for low emissions and economy and are not coveted by enthusiasts.

Beginning in 1986, the rear main oil seal design was changed from a two-piece to a one-piece seal. These late model blocks must have an oil seal retainer and use the late model crankshaft and flywheel/driveplate.

On 1987 and newer models with roller lifters, the block has extra long lifter bosses. These are the only blocks that can accept factory roller lifters.

There are two different block-mounted starter bolt patterns, straight and diagonal. Some blocks may not have the needed holes for the starter bolts. These can be drilled and tapped if they are missing.

Dipstick location also varies. Most pre-1980 engines have the dipstick on the driver's side. Some engines have the dipstick tube on the other side or mounted to the oil pan. Be sure the location is compatible with your application.

Big block V8s

Big block Chevrolet engines have been in production since 1965. The first ones built were 396 cubic inch versions. Later passenger car versions were made in 402, 427 and 454 cubic inch displacements.

Additionally, larger trucks were available with 366 and 427 c.i. V8s which have higher deck heights, four-ring pistons and, in some cases, gear-driven camshafts. However, many of the other components on these engines are interchangeable with their passenger-car counterparts.

The different displacements are obtained by combining several bores and strokes in various combinations:

Displacement table

Bore (inches)	Stroke (inches)	Displacement (cubic inches)
3.935	3.76	366
4.094	3.76	396
4.124	3.76	402
4.251	3.76	427
4.251	4.00	454

Camshafts

Camshaft specifications vary considerably between different model years and horsepower versions. Hydraulic, solid and roller lifter camshafts are available to tailor the engine to a specific application. Factory roller lifter camshafts must be used with blocks originally equipped

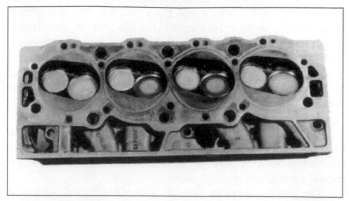

1.10 Closed chamber heads have more of a bathtub shape to the combustion chamber

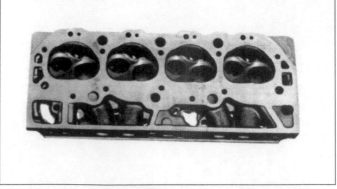

1.11 Open chamber heads have more of a "D" shape to the combustion chamber

with rollers. The camshaft specifications should closely match the rest of the engine to assure the best performance, driveability, economy and lowest emissions.

On some 366 and 427 engines used in large trucks, the camshaft is driven by gears instead of a chain and sprockets. These gear-driven models can be converted to chains to allow use of automotive grind camshafts. If this is done, the distributor driven gear must be changed because the gear-driven cams rotate in the opposite direction.

It is often difficult to determine which camshaft is installed in a particular used engine. Even after the camshaft is removed, casting numbers are unreliable for identifying specific grinds. Although used camshafts can be installed in another engine if the lifters are reinstalled on the same lobes, we recommend new camshafts and lifters when replacement is necessary. Camshafts and lifters should always be replaced together as a set.

Pushrods

Five different types of pushrods have been installed by the factory in big block Chevrolet engines. The primary differences between the various types are the tip designs and the diameter. **Note:** *Intake and exhaust valves use different length pushrods.*

Pushrods are available in three different diameters: 5/16, 3/8 and 7/16 inch. The strongest and preferred pushrods are no. 3942415 for the exhaust and no. 3942416 for the intake. Always use the matching guide plates for the pushrods.

Cylinder heads

There are two basic combustion chamber designs: closed chamber and open chamber. Closed chamber heads, which were used until 1971, can be identified by the characteristic "bathtub" shape and steep combustion chamber walls close to the valves **(see illustration)**.

Open chamber heads **(see illustration)** are preferred over closed chamber heads and they are used on all engines from 1972 on. These can be identified by the more triangular shape of the combustion chamber at the gasket surface.

The piston crowns must match the type of combustion

chamber. Failure to do so may result in pistons colliding with cylinder heads.

Cylinder heads for the big block Chevy were made in both aluminum and cast iron. Virtually all aluminum heads are designed for high performance use, and they are expensive and difficult to find. Some aluminum heads have extra bolt holes for additional clamping force. Always compare the gaskets to the heads and block to ensure compatibility.

Cast-iron heads are available in both standard small round port and high-performance rectangular large port versions, with open and closed chamber configurations and with various valve sizes and combustion chamber volumes. Always check carefully when comparing cylinder heads to obtain the correct ones.

Beginning with 1991 models, the rocker arms are installed with shoulder bolts and cast aluminum valve covers replace the stamped steel units.

Whenever possible, obtain replacement cylinder heads in matched pairs to ensure they are the same.

Intake manifolds

There are a great many choices in intake manifolds. The factory has offered two-barrel, four-barrel and three-two-barrel carburetor manifolds and fuel injection on the later models.

If you are working on a fuel-injected model or restoring a vehicle to original condition, an identical part must be obtained. We will not go into detail on fuel injected or multi-carb models since the choices are limited. However, for most applications using a single carburetor, several types of manifolds will work satisfactorily.

Some low performance engines were built with cast iron manifolds and two-barrel carburetors. However, a single four-barrel carb works best for most applications. These are available with mounting provisions for Rochester and Holley carburetors. Lower horsepower engines can use manifold part number 3977608 with a spread-bore Holley four-barrel. Many standard performance four-barrel models used Rochester Quadrajets.

Several good aluminum four-barrel manifolds were installed at the factory. Exchanging a cast-iron manifold for an aluminum one is an easy way to shave weight off the front end of a vehicle.

For high-performance street use, the factory aluminum high-rise is hard to beat. It mounts a Holley four-barrel and goes under part number 3947084.

Always compare the port sizes of the heads and manifolds by holding a gasket up to both openings. For optimum performance the sizes and shapes must match. Larger ports favor high rpm and larger displacements. Small ports tend to improve low rpm torque and driveability.

Medium and heavy-duty truck engines (366 and 427) with higher deck heights must use truck manifolds or require spacers to adapt passenger car manifolds to the heads. These truck manifolds cannot be used on passenger car engines.

On 1973 and later models, look for Exhaust Gas Recirculation (EGR) valve provisions. Always compare the old and new manifolds to determine if sufficient vacuum fittings are present.

Exhaust manifolds

Stock exhaust manifolds tend to be restrictive and limit power. The most efficient manifolds were installed on Corvettes. However, the Corvette manifolds won't fit in most other engine compartments. Look for the manifolds from the highest horsepower version originally available in the chassis/engine combination. Many enthusiasts convert to tubular headers if they are looking for more performance than this affords.

Heat risers or EFE valves, if used, must fit the manifold correctly. Also, accessory mounts and the length and angle of exhaust outlet must match the engine and vehicle. Be sure to check these factors before purchase.

Crankshafts

Crankshafts from all 366, 396, 402 and 427 engines have the same stroke and are interchangeable. Although the 427 crank is balanced differently than the others, it can be swapped if the engine is rebalanced.

Forged crankshafts are stronger and preferred for heavy-duty use. Factory high-performance crankshafts are forged and low-performance versions are cast. Cast cranks have a narrow parting line "flash" and forged cranks have a wider line **(see illustration)**.

The 454 crankshaft has a longer stroke than the other big blocks and is externally balanced. This crank (pre-1991) can be used in smaller displacement engines if the 454 pistons are used along with the 454 harmonic balancer and flywheel/driveplate.

Beginning with 1991 models, the two-piece rear crankshaft oil seal is replaced with a one-piece unit. This necessitated a change in crankshaft flywheel flange and block design. The 1991 and newer cast-iron crankshaft, block and flywheel/driveplate are thus not interchangeable with earlier models.

Blocks

The vast majority of blocks are cast iron. Only a few ZL-

1 racing engines were cast in aluminum, and most of these are in the hands of collectors.

Smaller displacement engines can be bored out and stock pistons installed to create 402, 427 and 454s from 396 versions. The correct stroke crankshaft must be used, and on the 396 minor grinding is required at the bottom of some cylinders for connecting rod clearance.

Four bolt main bearing caps are desirable for heavy duty use. Perhaps the easiest way to identify a four bolt main block without removing the oil pan is by checking the oil cooler fittings above the oil filter mount. If there is only one hole the engine is probably a two bolt main version. If there is one 1/2 inch pipe plug and one 3/4 inch pipe plug, the engine probably has four bolt mains.

Engine blocks used on 366 and 427 large truck engines are 0.4 inch taller than passenger car versions. Some, but not all truck blocks are marked "Truck" for identification. Most of the parts are interchangeable, except for the pistons, pushrods, gear drive camshaft assemblies, manifolds and distributors.

Beginning with 1986 models, the two-piece rear crankshaft oil seal is replaced with a one-piece unit. The 1986 and newer cast iron crankshaft and block are thus not interchangeable with earlier models.

Buying parts

Commonly replaced engine parts such as piston rings, pistons, bearings, camshafts, lifters, timing chains, oil pumps and gaskets are produced by aftermarket manufacturers and stocked by retail auto parts stores and mail order houses, usually at a savings over dealer parts department prices. Many auto parts stores and mail order houses offer complete engine kits, often at a considerable savings over individual parts (see *Rebuilding options* in Chapter 3 for more information). Don't buy gaskets separately. A good-quality complete engine overhaul gasket set (available at

1.12 This is a cast-iron crankshaft; forged crankshafts have a wider parting line "flash" (A) - the casting number (B) can tell you still more about the crankshaft

most auto parts stores) will save you money and the needless hassle of buying individual gaskets.

Less-commonly replaced items such as fuel pump drives and rocker arm shafts may not be available through these same sources and a dealer service department may be your only option. Keep in mind that some parts will probably have to be ordered, and it may take several days to get your parts; order early.

Wrecking yards are a good source for major parts that would otherwise only be available through a dealer service department (where the price would likely be high). Engine blocks, cylinder heads, crankshafts, manifolds, etc. for these engines are commonly available for reasonable prices. The parts people at wrecking yards have parts interchange books they can use to quickly identify parts from other models and years that are the same as the ones on your engine. If you have a high-performance version of an engine, don't expect to find high-performance parts easily; they're usually snatched up quickly from wrecking yards by savvy scroungers.

2 Tools and equipment

A place to work

Establish a place to work. A special work area is essential. It doesn't have to be particularly large, but it should be clean, safe, well-lit, organized and adequately equipped for the job. True, without a good workshop or garage, you can still service and repair engines, even if you have to work outside. But an overhaul or major repairs should be carried out in a sheltered area with a roof. The procedures in this book require an environment totally free of dirt, which will cause wear if it finds its way into an engine.

The workshop

The size, shape and location of a shop building is usually dictated by circumstance rather than personal choice. Every do-it-yourselfer dreams of having a spacious, clean, well-lit building specially designed and equipped for working on everything from small engines on lawn and garden equipment to cars and other vehicles. In reality, however, most of us must content ourselves with a garage, basement or shed in the backyard.

Spend some time considering the potential - and drawbacks - of your current facility. Even a well-established workshop can benefit from intelligent design. Lack of space is the most common problem, but you can significantly increase usable space by carefully planning the locations of work and storage areas. One strategy is to look at how others do it. Ask local repair shop owners if you can see their shops. Note how they've arranged their work areas, storage and lighting, then try to scale down their solutions to fit your own shop space, finances and needs.

General workshop requirements

A solid concrete floor is the best surface for a shop area. The floor should be even, smooth and dry. A coat of paint or sealant formulated for concrete surfaces will make oil spills and dirt easier to remove and help cut down on dust - always a problem with concrete.

Paint the walls and ceiling white for maximum reflection. Use gloss or semi-gloss enamel. It's washable and reflective. If your shop has windows, situate workbenches to take advantage of them. Skylights are even better. You can't have too much natural light. Artificial light is also good, but you'll need a lot of it to equal ordinary daylight.

Make sure the building is adequately ventilated. This is critical during the winter months, to prevent condensation problems. It's also a vital safety consideration where solvents, gasoline and other volatile liquids are being used. You should be able to open one or more windows for ventilation. In addition, opening vents in the walls are desirable.

Electricity and lights

Electricity is essential in a shop. It's relatively easy to install if the workshop is part of the house, but it can be difficult and expensive to install if it isn't. Safety should be your primary consideration when dealing with electricity; unless you have a very good working knowledge of electrical installations, have an electrician do any work required to provide power and lights in the shop.

Consider the total electrical requirements of the shop, making allowances for possible later additions of lights and equipment. Don't substitute extension cords for legal and safe permanent wiring. If the wiring isn't adequate, or is

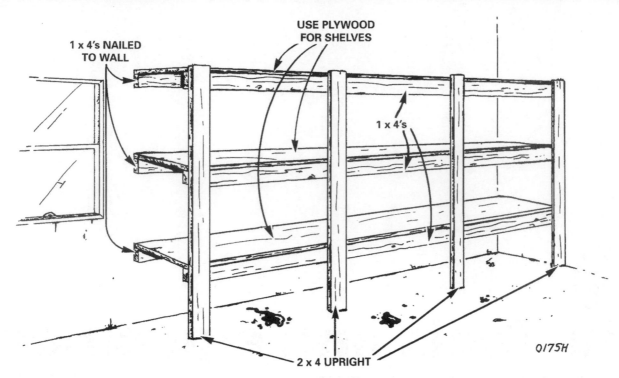

1 x 4's NAILED
TO WALL

USE PLYWOOD
FOR SHELVES

1 x 4's

2 x 4 UPRIGHT

Q175H

2.1 Homemade wood shelves are relatively inexpensive to build and you can design them to fit the available space, but all that wood can be a fire hazard

substandard, have it upgraded.

Give careful consideration to lights for the workshop. A pair of 150-watt incandescent bulbs, or two 48-inch long, 40-watt fluorescent tubes, suspended approximately 48-inches above the workbench, are the minimum you can get by with. As a general rule, fluorescent lights are probably the best choice. Their light is bright, even, shadow-free and fairly economical, although some people don't care for the bluish tinge they cast on everything. The usual compromise is a good mix of fluorescent and incandescent fixtures.

The position of the lights is important. Don't place a fixture directly above the area where the engine - or the stand it's mounted on - is located. It will cause shadows, even with fluorescent lights. Attach the light(s) slightly to the rear - or to each side - of the workbench or engine stand to provide shadow-free lighting. A portable "trouble-light" is very helpful for use when overhead lights are inadequate. If gasoline, solvents or other flammable liquids are present - not an unusual situation in a shop - use special fittings to minimize the risk of fire. And don't use fluorescent lights above machine tools (like a drill press). The flicker produced by alternating current is especially pronounced with this type of light and can make a rotating chuck appear stationary at certain speeds - a very dangerous situation.

Storage and shelves

Once disassembled, an engine occupies more space than you might think. Set up an organized storage area to avoid losing parts. You'll also need storage space for hard-

ware, lubricants, solvent, rags, tools and equipment.

If space and finances allow, install metal shelves along the walls. Arrange the shelves so they're widely spaced near the bottom to take large or heavy items. Metal shelf units are pricey, but they make the best use of available space. And the shelf height is adjustable on most units.

Wood shelves **(see illustration)** are sometimes a cheaper storage solution. But they must be built - not just assembled. They must be much heftier than metal shelves to carry the same weight, the shelves can't be adjusted vertically and you can't just disassemble them and take them with you if you move. Wood also absorbs oil and other liquids and is obviously a much greater fire hazard.

Store small parts in plastic drawers or bins mounted on metal racks attached to the wall. They're available from most hardware, home and lumber stores. Bins come in various sizes and usually have slots for labels.

All kinds of containers are useful in a shop. Glass jars are handy for storing fasteners, but they're easily broken. Cardboard boxes are adequate for temporary use, but if they become damp, the bottoms eventually weaken and fall apart if you store oily or heavy parts in them. Plastic containers come in a variety of sizes and colors for easy identification. Egg cartons are excellent organizers for tiny parts like valve springs, retainers and keepers. Large ice cream tubs are suitable for keeping small parts together. Get the type with a snap cover. Old metal cake pans, bread pans and muffin tins also make good storage containers for small parts.

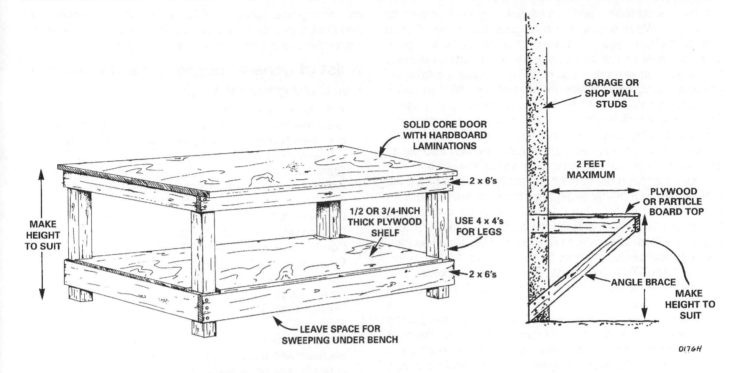

2.2 You can build a sturdy, inexpensive workbench with 4 X 4s, 2 X 6s and a solid core door with hardboard laminations - or build a bench using the wall as an integral member as shown

Workbenches

A workbench is essential - it provides a place to lay out parts and tools during repair procedures, and it's a lot more comfortable than working on a floor or the driveway. The workbench should be as large and sturdy as space and finances allow. If cost is no object, buy industrial steel benches. They're more expensive than home-built benches, but they're very strong, they're easy to assemble, and - if you move - they can be disassembled quickly and you can take them with you. They're also available in various lengths, so you can buy the exact size to fill the space along a wall.

If steel benches aren't in the budget, fabricate a bench frame from slotted angle-iron or Douglas fir (use 2 x 6's rather than 2 x 4's) **(see illustration)**. Cut the pieces of the frame to the required size and bolt them together with carriage bolts. A 30 or 36 by 80-inch, solid-core door with hardboard surfaces makes a good bench top. And you can flip it over when one side is worn out.

An even cheaper - and quicker - solution? Assemble a bench by attaching the bench top frame pieces to the wall with angled braces and use the wall studs as part of the framework.

Regardless of the type of frame you decide to use for the workbench, be sure to position the bench top at a comfortable working height and make sure everything is level. Shelves installed below the bench will make it more rigid and provide useful storage space.

Tools and equipment

For some home mechanics, the idea of using the correct tool is completely foreign. They'll cheerfully tackle the most complex overhaul procedures with only a set of cheap open-end wrenches of the wrong type, a single screwdriver with a worn tip, a large hammer and an adjustable wrench. Though they often get away with it, this cavalier approach is stupid and dangerous. It can result in relatively minor annoyances like stripped fasteners, or cause catastrophic consequences like blown engines. It can also result in serious injury.

A complete assortment of good tools is a given for anyone who plans to overhaul engines. If you don't already have most of the tools listed below, the initial investment may seem high, but compared to the spiraling costs of routine maintenance and repairs, it's a deal. Besides, you can use a lot of the tools around the house for other types of mechanical repairs. We've included a list of the tools you'll need and a detailed description of what to look for when shopping for tools and how to use them correctly. We've also included a list of the special factory tools you'll need for engine rebuilding.

Buying tools

There are two ways to buy tools. The easiest and quickest way is to simply buy an entire set. Tool sets are often priced substantially below the cost of the same indi-

Haynes Chevrolet engine overhaul manual

vidually priced tools - and sometimes they even come with a tool box. When purchasing such sets, you often wind up with some tools you don't need or want. But if low price and convenience are your concerns, this might be the way to go. Keep in mind that you're going to keep a quality set of tools a long time (maybe the rest of your life), so check the tools carefully; don't skimp too much on price, either. Buying tools individually is usually a more expensive and time-consuming way to go, but you're more likely to wind up with the tools you need and want. You can also select each tool on its relative merits for the way you use it.

You can get most of the hand tools on our list from the tool department of any large department store or hardware store chain that sells hand tools. Blackhawk, Cornwall, Craftsman, KD, Proto and SK are fairly inexpensive, good-quality choices. Specialty tools are available from mechanics' tool companies such as Snap-on, Mac, Matco, Kent-Moore, Lisle, OTC, Owatonna, etc. These companies also supply the other tools you need, but they'll probably be more expensive.

Also consider buying second-hand tools from garage sales or used tool outlets. You may have limited choice in sizes, but you can usually determine from the condition of the tools if they're worth buying. You can end up with a number of unwanted or duplicate tools, but it's a cheap way of putting a basic tool kit together, and you can always sell off any surplus tools later.

Until you're a good judge of the quality levels of tools, avoid mail order firms (excepting Sears and other name-brand suppliers), flea markets and swap meets. Some of them offer good value for the money, but many sell cheap, imported tools of dubious quality. Like other consumer products counterfeited in the Far East, these tools run the gamut from acceptable to unusable.

If you're unsure about how much use a tool will get, the following approach may help. For example, if you need a set of combination wrenches but aren't sure which sizes you'll end up using most, buy a cheap or medium-priced set (make sure the jaws fit the fastener sizes marked on them). After some use over a period of time, carefully examine each tool in the set to assess its condition. If all the tools fit well and are undamaged, don't bother buying a better set. If one or two are worn, replace them with high-quality items - this way you'll end up with top-quality tools where they're needed most and the cheaper ones are sufficient for occasional use. On rare occasions you may conclude the whole set is poor quality. If so, buy a better set, if necessary, and remember never to buy that brand again.

In summary, try to avoid cheap tools, especially when you're purchasing high-use items like screwdrivers, wrenches and sockets. Cheap tools don't last long. Their initial cost plus the additional expense of replacing them will exceed the initial cost of better-quality tools.

Hand tools

Note: *The information that follows is for early-model engines with only Standard fastener sizes. On some late-* model engines, you'll need Metric wrenches, sockets and Allen wrenches. Generally, manufacturers began integrating metric fasteners into their vehicles around 1975.

A list of general-purpose hand tools you need for general engine work

Adjustable wrench - 10-inch
Allen wrench set (1/8 to 3/8-inch or 4 mm to 10 mm)
Ball peen hammer - 12 oz (any steel hammer will do)
Box-end wrenches
Brass hammer
Brushes (various sizes, for cleaning small passages
Combination (slip-joint) pliers - 6-inch
Center punch
Cold chisels - 1/4 and 1/2-inch
Combination wrench set (1/4 to 1-inch)
Extensions - 1-, 6-, 10- and 12-inch
E-Z out (screw extractor) set
Feeler gauge set
Files (assorted)
Floor jack
Gasket scraper
Hacksaw and assortment of blades
Impact screwdriver and bits
Locking pliers
Micrometer(s) (one-inch)
Phillips screwdriver (no. 2 x 6-inch)
Phillips screwdriver (no. 3 x 8-inch)
Phillips screwdriver (stubby - no. 2)
Pin punches (1/16, 1/8, 3/16-inch)
Piston ring removal and installation tool
Pliers - lineman's
Pliers - needle-nose
Pliers - snap-ring (internal and external)
Pliers - vise-grip
Pliers - diagonal cutters
Ratchet (reversible)
Scraper (made from flattened copper tubing)
Scribe
Socket set (6-point)
Soft-face hammer (plastic/rubber, the biggest
   you can buy)
Spark plug socket (with rubber insert)
Spark plug gap adjusting tool
Standard screwdriver (1/4-inch x 6-inch)
Standard screwdriver (5/16-inch x 6-inch)
Standard screwdriver (3/8-inch x 10-inch)
Standard screwdriver (5/16-inch - stubby)
Steel ruler - 6-inch
Straightedge - 12-inch
Tap and die set
Thread gauge
Torque wrench (same size drive as sockets)
Torx socket(s)
Universal joint
Wire brush (large)
Wire cutter pliers

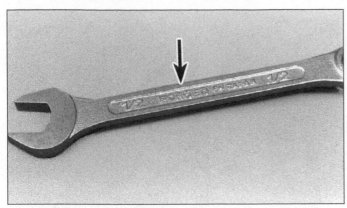

2.3 One quick way to determine whether you're looking at a quality wrench is to read the information printed on the handle - if it says, "chrome vanadium" or "forged," it's made out of the right material

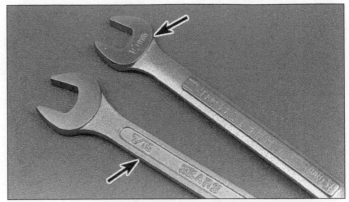

2.4 The size stamped on a wrench indicates the distance across the nut or bolt head (or the distance between the wrench jaws) in inches, not the diameter of the threads on the fastener

What to look for when buying hand tools and general purpose tools

Wrenches and sockets

Wrenches vary widely in quality. One indication of their cost is their quality: The more they cost, the better they are. Buy the best wrenches you can afford. You'll use them a lot.

Start with a set containing wrenches from 1/4 to 1-inch in size. The size, stamped on the wrench **(see illustration)**, indicates the distance across the nut or bolt head, or the distance between the wrench jaws - not the diameter of the threads on the fastener - in inches. For example, a 1/4-inch bolt usually has a 7/16-inch hex head - the size of the wrench required to loosen or tighten it. However, the relationship between thread diameter and hex size doesn't always hold true. In some instances, an unusually small hex may be used to discourage over-tightening or because space around the fastener head is limited. Conversely, some fasteners have a disproportionately large hex-head.

Wrenches are similar in appearance, so their quality level can be difficult to judge just by looking at them. There are bargains to be had, just as there are overpriced tools with well-known brand names. On the other hand, you may buy what looks like a reasonable value set of wrenches only to find they fit badly or are made from poor-quality steel.

With a little experience, it's possible to judge the quality of a tool by looking at it. Often, you may have come across the brand name before and have a good idea of the quality. Close examination of the tool can often reveal some hints as to its quality. Prestige tools are usually polished and chrome-plated over their entire surface, with the working faces ground to size. The polished finish is largely cosmetic, but it does make them easy to keep clean. Ground jaws normally indicate the tool will fit well on fasteners. A side-by-side comparison of a high-quality wrench with a cheap equivalent is an eye opener. The better tool will be made from a good-quality material, often a forged/chrome-vanadium steel alloy **(see illustration)**. This, together with careful design, allows the tool to be kept as small and compact as possible. If, by comparison, the cheap tool is

thicker and heavier, especially around the jaws, it's usually because the extra material is needed to compensate for its lower quality. If the tool fits properly, this isn't necessarily bad - it is, after all, cheaper - but in situations where it's necessary to work in a confined area, the cheaper tool may be too bulky to fit.

Open-end wrenches

Because of its versatility, the open-end wrench is the most common type of wrench. It has a jaw on either end, connected by a flat handle section. The jaws either vary by a size, or overlap sizes between consecutive wrenches in a set. This allows one wrench to be used to hold a bolt head while a similar-size nut is removed. A typical fractional size wrench set might have the following jaw sizes: 1/4 x 5/16, 3/8 x 7/16, 1/2 x 9/16, 9/16 x 5/8 and so on.

Typically, the jaw end is set at an angle to the handle, a feature which makes them very useful in confined spaces; by turning the nut or bolt as far as the obstruction allows, then turning the wrench over so the jaw faces in the other direction, it's possible to move the fastener a fraction of a turn at a time (see illustration). The handle length is gener-

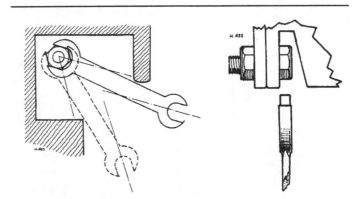

2.5 Open-end wrenches can do several things other wrenches can't - for example, they can be used on bolt heads with limited clearance (above) and they can be used in tight spots where there's little room to turn a wrench by flipping the offset jaw over every few degrees of rotation

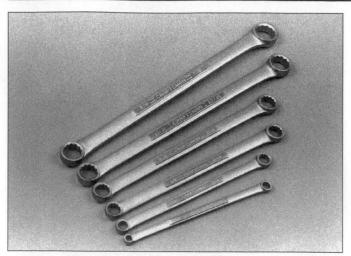

2.6 Box-end wrenches have a ring-shaped "box" at each end - when space permits, they offer the best combination of "grip" and strength

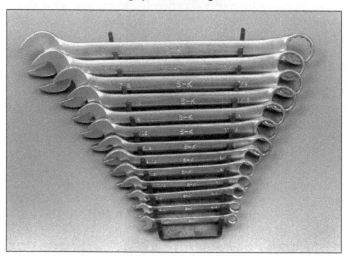

2.8 Buy a set of combination wrenches from 1/4 to 1-inch

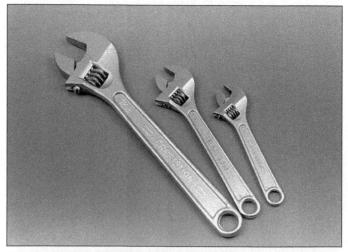

2.9 Adjustable wrenches can handle a range of fastener sizes - they're not as good as single-size wrenches, but they're handy for loosening and tightening those odd-sized fasteners for which you haven't yet bought the correct wrench

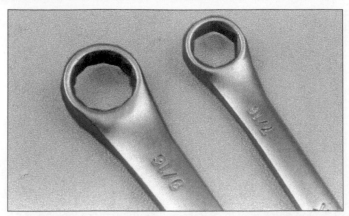

2.7 Box-end wrenches are available in 12- (left) and 6-point (right) openings; even though the 12-point design offers twice as many wrench positions, but the 6-point first - it's less likely to strip off the corners of a nut or bolt head

ally determined by the size of the jaw and is calculated to allow a nut or bolt to be tightened sufficiently by hand with minimal risk of breakage or thread damage (though this doesn't apply to soft materials like brass or aluminum).

Common open-end wrenches are usually sold in sets and it's rarely worth buying them individually unless it's to replace a lost or broken tool from a set. Single tools invariably cost more, so check the sizes you're most likely to need regularly and buy the best set of wrenches you can afford in that range of sizes. If money is limited, remember that you'll use open-end wrenches more than any other type - it's a good idea to buy a good set and cut corners elsewhere.

Box-end wrenches

Box-end wrenches **(see illustration)** have ring-shaped ends with a 6-point (hex) or 12-point (double hex) opening **(see illustration)**. This allows the tool to fit on the fastener hex at 15 (12-point) or 30-degree (6-point) intervals. Normally, each tool has two ends of different sizes, allowing an overlapping range of sizes in a set, as described for open-end wrenches.

Although available as flat tools, the handle is usually offset at each end to allow it to clear obstructions near the fastener, which is normally an advantage. In addition to normal length wrenches, it's also possible to buy long handle types to allow more leverage (very useful when trying to loosen rusted or seized nuts). It is, however, easy to shear off fasteners if not careful, and sometimes the extra length impairs access.

As with open-end wrenches, box-ends are available in varying quality, again often indicated by finish and the amount of metal around the ring ends. While the same criteria should be applied when selecting a set of box-end wrenches, if your budget is limited, go for better-quality open-end wrenches and a slightly cheaper set of box-ends.

Combination wrenches

These wrenches **(see illustration)** combine a box-end and open-end of the same size in one tool and offer many

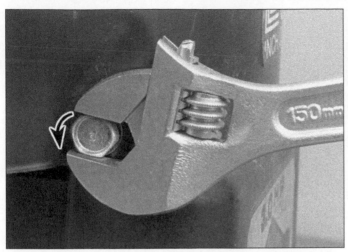

2.10 When you use an adjustable wrench, make sure the movable jaw points in the direction the wrench is being turned (arrow) so the wrench doesn't distort and slip off the fastener head

of the advantages of both. Like the others, they're widely available in sets and as such are probably a better choice than box-ends only. They're generally compact, short-handled tools and are well suited for tight spaces where access is limited.

Adjustable wrenches

Adjustable wrenches **(see illustration)** come in several sizes. Each size can handle a range of fastener sizes. Adjustable wrenches aren't as effective as one-size tools and it's easy to damage fasteners with them. However, they can be an invaluable addition to any tool kit - if they're used with discretion. **Note:** *If you attach the wrench to the fastener with the movable jaw pointing in the direction of wrench rotation* **(see illustration)**, *an adjustable wrench will be less likely to slip and damage the fastener head.*

The most common adjustable wrench is the open-end type with a set of parallel jaws that can be set to fit the head of a fastener. Most are controlled by a threaded spindle, though there are various cam and spring-loaded versions available. Don't buy large tools of this type; you'll rarely be able to find enough clearance to use them.

Ratchet and socket sets

Ratcheting socket wrenches **(see illustration)** are highly versatile. Besides the sockets themselves, many other interchangeable accessories - extensions, U-drives, step-down adapters, screwdriver bits, Allen bits, crow's feet, etc. - are available. Buy six-point sockets - they're less likely to slip and strip the corners off bolts and nuts. Don't buy sockets with extra-thick walls - they might be stronger but they can be hard to use on recessed fasteners or fasteners in tight quarters.

Buy a 3/8-inch drive for work on the outside of the engine. It's the one you'll use most of the time. Get a 1/2-inch drive for overhaul work. Although the larger drive is bulky and more expensive, it has the capacity of accepting a very wide range of large sockets. Later, you may want to

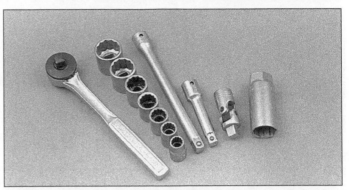

2.11 A typical ratchet and socket set includes a ratchet, a set of sockets, a long and a short extension, a universal joint and a spark plug socket

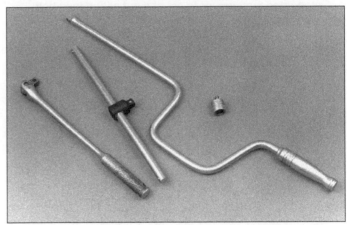

2.12 Lots of other accessories are available for ratchets: From left to right, a breaker bar, a sliding T-handle, a speed handle and a 3/8-to-1/4-inch adapter

consider a 1/4-inch drive for little stuff like ignition and carburetor work.

Interchangeable sockets consist of a forged-steel alloy cylinder with a hex or double-hex formed inside one end. The other end is formed into the square drive recess that engages over the corresponding square end of various socket drive tools.

Sockets are available in 1/4, 3/8, 1/2 and 3/4-inch drive sizes. A 3/8-inch drive set is most useful for engine repairs, although 1/4-inch drive sockets and accessories may occasionally be needed.

The most economical way to buy sockets is in a set. As always, quality will govern the cost of the tools. Once again, the "buy the best" approach is usually advised when selecting sockets. While this is a good idea, since the end result is a set of quality tools that should last a lifetime, the cost is so high it's difficult to justify the expense for home use.

As far as accessories go, you'll need a ratchet, at least one extension (buy a three or six-inch size), a spark plug socket and maybe a T-handle or breaker bar. Other desirable, though less essential items, are a speeder handle, a U-joint, extensions of various other lengths and adapters from one drive size to another **(see illustration)**. Some of the sets you find may combine drive sizes; they're well

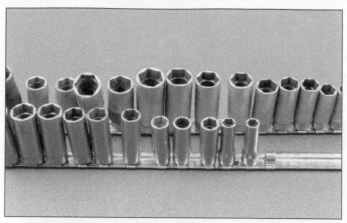

2.13 Deep sockets enable you to loosen or tighten an elongated fastener, or to get at a nut with a long bolt protruding from it

2.14 Standard and Phillips bits, Allen-head and Torx drivers will expand the versatility of your ratchet and extensions even further

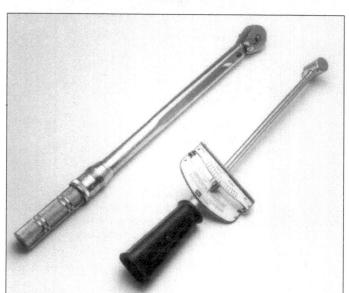

2.15 Torque wrenches (click-type on left, beam-type on right) are the only way to accurately tighten critical fasteners like connecting rod bolts, cylinder head bolts, etc.

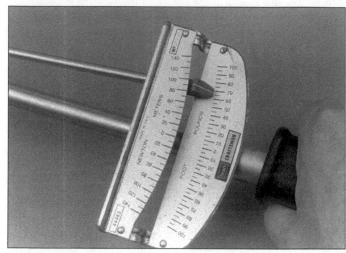

2.16 The deflecting beam-type torque wrench is inexpensive and simple to use - just tighten the fastener until the pointer points to the specified torque setting

worth having if you find the right set at a good price, but avoid being dazzled by the number of pieces.

Above all, be sure to completely ignore any label that reads "86-piece Socket Set," which refers to the number of pieces, not to the number of sockets (sometimes even the metal box and plastic insert are counted in the total!).

Apart from well-known and respected brand names, you'll have to take a chance on the quality of the set you buy. If you know someone who has a set that has held up well, try to find the same brand, if possible. Take a pocketful of nuts and bolts with you and check the fit in some of the sockets. Check the operation of the ratchet. Good ones operate smoothly and crisply in small steps; cheap ones are coarse and stiff - a good basis for guessing the quality of the rest of the pieces.

One of the best things about a socket set is the built-in facility for expansion. Once you have a basic set, you can purchase extra sockets when necessary and replace worn or damaged tools. There are special deep sockets for reaching recessed fasteners or to allow the socket to fit over a projecting bolt or stud **(see illustration)**. You can also buy screwdriver, Allen and Torx bits to fit various drive tools (they can be very handy in some applications) **(see illustration)**. Most socket sets include a special deep socket for 14 millimeter spark plugs. They have rubber inserts to protect the spark plug porcelain insulator and hold the plug in the socket to avoid burned fingers.

Torque wrenches

Torque wrenches **(see illustration)** are essential for tightening critical fasteners like rod bolts, main bearing cap bolts, head bolts, etc. Attempting an engine overhaul without a torque wrench is an invitation to oil leaks, distortion of the cylinder head, damaged or stripped threads or worse.

There are two types of torque wrenches - the "beam" type, which indicates torque loads by deflecting a flexible shaft and the "click" type **(see illustrations)**, which emits

2.17 "Click" type torque wrenches can be set to "give" at a pre-set torque, which makes them very accurate and easy to use

2.18 The impact driver converts a sharp blow into a twisting motion - this is a handy addition to your socket arsenal for those fasteners that won't let go - you can use it with any bit that fits a 3/8-inch drive ratchet

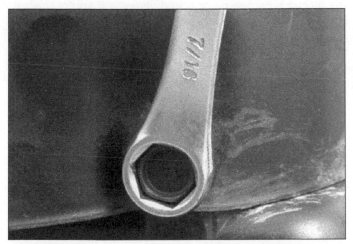

2.19 Try to use a six-point box wrench (or socket) whenever possible - its shape matches that of the fastener, which means maximum grip and minimum slip

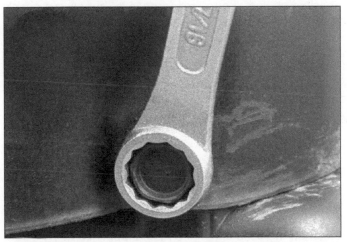

2.20 Sometimes a six-point tool just doesn't offer you any grip when you get the wrench at the angle it needs to be in to loosen or tighten a fastener - when this happens, pull out the 12-point sockets or wrenches - but remember: they're much more likely to strip the corners off a fastener

an audible click when the torque resistance reaches the specified resistance.

Torque wrenches are available in a variety of drive sizes and torque ranges for particular applications. For engine rebuilding, 0 to 150 ft-lbs should be adequate. Keep in mind that "click" types are usually more accurate (and more expensive).

Impact drivers

The impact driver **(see illustration)** belongs with the screwdrivers, but it's mentioned here since it can also be used with sockets (impact drivers normally are 3/8-inch square drive). As explained later, an impact driver works by converting a hammer blow on the end of its handle into a sharp twisting movement. While this is a great way to jar a seized fastener loose, the loads imposed on the socket are excessive. Use sockets only with discretion and expect to have to replace damaged ones on occasion.

Using wrenches and sockets

Although you may think the proper use of tools is self-evident, it's worth some thought. After all, when did you last see instructions for use supplied with a set of wrenches?

Which wrench?

Before you start tearing an engine apart, figure out the best tool for the job; in this instance the best wrench for a hex-head fastener. Sit down with a few nuts and bolts and look at how various tools fit the bolt heads.

A golden rule is to choose a tool that contacts the largest area of the hex-head. This distributes the load as evenly as possible and lessens the risk of damage. The shape most closely resembling the bolt head or nut is another hex, so a 6-point socket or box-end wrench is usually the best choice **(see illustration)**. Many sockets and box-end wrenches have double hex (12-point) openings. If you slip a 12-point box-end wrench over a nut, look at how and where the two are in contact. The corners of the nut engage in every other point of the wrench. When the wrench is turned, pressure is applied evenly on each of the six corners **(see illustration)**. This is fine unless the fastener head was previously rounded off. If so, the corners will be damaged and the wrench will slip. If you encounter a damaged bolt head or nut, always use a 6-point wrench or socket if possible. If you don't have one of the right size, choose a wrench that fits securely and proceed with care.

If you slip an open-end wrench over a hex-head fastener, you'll see the tool is in contact on two faces only **(see illustration)**. This is acceptable provided the tool and fastener are both in good condition. The need for a snug fit between the wrench and nut or bolt explains the recommendation to buy good-quality open-end wrenches. If the wrench jaws, the bolt head or both are damaged, the wrench will probably slip, rounding off and distorting the head. In some applications, an open-end wrench is the only possible choice due to limited access, but always check the fit of the wrench on the fastener before attempting to loosen it; if it's hard to get at with a wrench, think how hard it will be to remove after the head is damaged.

The last choice is an adjustable wrench or self-locking plier/wrench (Vise-Grips). Use these tools only when all else has failed. In some cases, a self-locking wrench may be able to grip a damaged head that no wrench could deal with, but be careful not to make matters worse by damaging it further.

Bearing in mind the remarks about the correct choice of tool in the first place, there are several things worth noting about the actual use of the tool. First, make sure the wrench head is clean and undamaged. If the fastener is rusted or coated with paint, the wrench won't fit correctly. Clean off the head and, if it's rusted, apply some penetrating oil. Leave it to soak in for a while before attempting removal.

It may seem obvious, but take a close look at the fastener to be removed before using a wrench. On many mass-produced machines, one end of a fastener may be fixed or captive, which speeds up initial assembly and usually makes removal easier. If a nut is installed on a stud or a bolt threads into a captive nut or tapped hole, you may have only one fastener to deal with. If, on the other hand, you have a separate nut and bolt, you must hold the bolt head while the nut is removed. In some areas this can be difficult, particularly where engine mounts are involved. In this type of situation you may need an assistant to hold the bolt head with a wrench while you remove the nut from the other side. If this isn't possible, you'll have to try to position a box-end wrench so it wedges against some other component to prevent it from turning.

Be on the lookout for left-hand threads. They aren't common, but are sometimes used on the ends of rotating shafts to make sure the nut doesn't come loose during engine operation (most engines covered by this book don't have these types of fasteners). If you can see the shaft end,

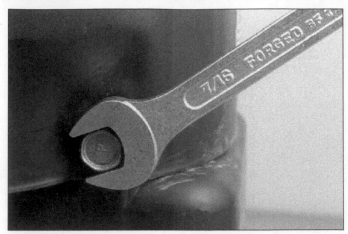

2.21 Open-end wrenches contact only two sides of the fastener and the jaws tend to open up when you put some muscle on the wrench handle - that's why they should only be used as a last resort

the thread type can be checked visually. If you're unsure, place your thumbnail in the threads and see which way you have to turn your hand so your nail "unscrews" from the shaft. If you have to turn your hand counterclockwise, it's a conventional right-hand thread.

Beware of the upside-down fastener syndrome. If you're loosening a fastener from the under side of a something, it's easy to get confused about which way to turn it. What seems like counterclockwise to you can easily be clockwise (from the fastener's point of view). Even after years of experience, this can still catch you once in a while. In most cases, a fastener can be removed simply by placing the wrench on the nut or bolt head and turning it. Occasionally, though, the condition or location of the fastener may make things more difficult. Make sure the wrench is square on the head. You may need to reposition the tool or try another type to obtain a snug fit. Make sure the engine you're working on is secure and can't move when you turn the wrench. If necessary, get someone to help steady it for you. Position yourself so you can get maximum leverage on the wrench.

If possible, locate the wrench so you can pull the end towards you. If you have to push on the tool, remember that it may slip, or the fastener may move suddenly. For this reason, don't curl your fingers around the handle or you may crush or bruise them when the fastener moves; keep your hand flat, pushing on the wrench with the heel of your thumb. If the tool digs into your hand, place a rag between it and your hand or wear a heavy glove.

If the fastener doesn't move with normal hand pressure, stop and try to figure out why before the fastener or wrench is damaged or you hurt yourself. Stuck fasteners may require penetrating oil, heat or an impact driver or air tool.

Using sockets to remove hex-head fasteners is less likely to result in damage than if a wrench is used. Make sure the socket fits snugly over the fastener head, then attach an extension, if needed, and the ratchet or breaker bar. Theoretically, a ratchet shouldn't be used for loosening

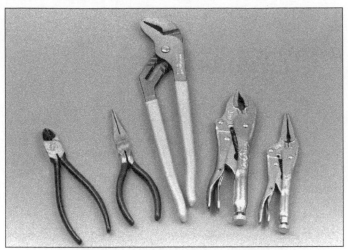

2.22 A typical assortment of the types of pliers you need to have in your box - from the left: diagonal cutters (dikes), needle-nose pliers, channel lock pliers, vise-grip pliers, needle-nose vise-grip pliers

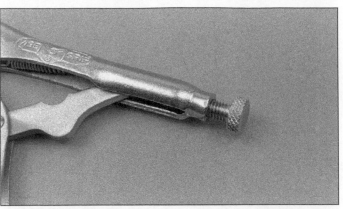

2.23 To adjust the jaws on a pair of vise-grips, grasp the part you want to hold with the jaws, tighten them down by turning the knurled knob on the end of one handle and snap the handles together - if you tightened the knob all the way down, you'll probably have to open it up (back it off) a little before you can close the handles

2.24 If you're persistent and careful, most fasteners can be removed with vise-grips

a fastener or for final tightening because the ratchet mechanism may be overloaded and could slip. In some instances, the location of the fastener may mean you have no choice but to use a ratchet, in which case you'll have to be extra careful.

Never use extensions where they aren't needed. Whether or not an extension is used, always support the drive end of the breaker bar with one hand while turning it with the other. Once the fastener is loose, the ratchet can be used to speed up removal.

Pliers

Some tool manufacturers make 25 or 30 different types of pliers. You only need a fraction of this selection **(see illustration)**. Get a good pair of slip-joint pliers for general use. A pair of needle-nose models is handy for reaching into hard-to-get-at places. A set of diagonal wire cutters (dikes) is essential for electrical work and pulling out cotter pins. Vise-Grips are adjustable, locking pliers that grip a fastener firmly - and won't let go - when locked into place. Parallel-jaw, adjustable pliers have angled jaws that remain parallel at any degree of opening. They're also referred to as Channel-lock (the original manufacturer) pliers, arc-joint pliers and water pump pliers. Whatever you call them, they're terrific for gripping a big fastener with a lot of force.

Slip-joint pliers have two open positions; a figure eight-shaped, elongated slot in one handle slips back-and-forth on a pivot pin on the other handle to change them. Good-quality pliers have jaws made of tempered steel and there's usually a wire-cutter at the base of the jaws. The primary uses of slip-joint pliers are for holding objects, bending and cutting throttle wires and crimping and bending metal parts, not loosening nuts and bolts.

Arc-joint or "Channel-lock" pliers have parallel jaws you can open to various widths by engaging different tongues and grooves, or channels, near the pivot pin. Since the tool expands to fit many size objects, it has countless uses for

engine and equipment maintenance. Channel-lock pliers come in various sizes. The medium size is adequate for general work; small and large sizes are nice to have as your budget permits. You'll use all three sizes frequently.

Vise-Grips (a brand name) come in various sizes; the medium size with curved jaws is best for all-around work. However, buy a large and small one if possible, since they're often used in pairs. Although this tool falls somewhere between an adjustable wrench, a pair of pliers and a portable vise, it can be invaluable for loosening and tightening fasteners - it's the only pliers that should be used for this purpose.

The jaw opening is set by turning a knurled knob at the end of one handle. The jaws are placed over the head of the fastener and the handles are squeezed together, locking the tool onto the fastener **(see illustration)**. The design of the tool allows extreme pressure to be applied at the jaws and a variety of jaw designs enable the tool to grip firmly even on damaged heads **(see illustration)**. Vise-Grips are great for removing fasteners that've been rounded off by badly-fitting wrenches.

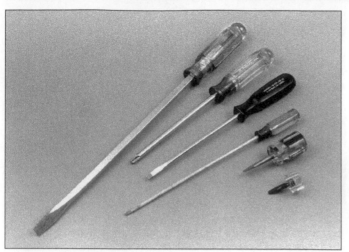

2.25 Screwdrivers come in myriad lengths, sizes and styles

Misuse of a screwdriver – the blade shown is both too narrow and too thin and will probably slip or break off

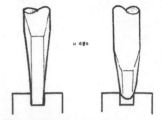

The left-hand example shows a snug-fitting tip. The right-hand drawing shows a damaged tip which will twist out of the slot when pressure is applied

2.26 Standard screwdrivers - wrong size (left), correct fit in screw slot (center) and worn tip (right)

As the name suggests, needle-nose pliers have long, thin jaws designed for reaching into holes and other restricted areas. Most needle-nose, or long-nose, pliers also have wire cutters at the base of the jaws.

Look for these qualities when buying pliers: Smooth operating handles and jaws, jaws that match up and grip evenly when the handles are closed, a nice finish and the word "forged" somewhere on the tool.

Screwdrivers

Screwdrivers **(see illustration)** come in a wide variety of sizes and price ranges. Anything from Craftsman on up is fine. But don't buy screwdriver sets for ten bucks at discount tool stores. Even if they look exactly like more expensive brands, the metal tips and shafts are made with inferior alloys and aren't properly heat treated. They usually bend the first time you apply some serious torque.

A screwdriver consists of a steel blade or shank with a drive tip formed at one end. The most common tips are standard (also called straight slot and flat-blade) and Phillips. The other end has a handle attached to it. Traditionally, handles were made from wood and secured to the shank, which had raised tangs to prevent it from turning in the handle. Most screwdrivers now come with plastic handles, which are generally more durable than wood.

The design and size of handles and blades vary considerably. Some handles are specially shaped to fit the human hand and provide a better grip. The shank may be either round or square and some have a hex-shaped bolster under the handle to accept a wrench to provide more leverage when trying to turn a stubborn screw. The shank diameter, tip size and overall length vary too. As a general rule, it's a good idea to use the longest screwdriver possible, which allows the greatest possible leverage.

If access is restricted, a number of special screwdrivers are designed to fit into confined spaces. The "stubby" screwdriver has a specially shortened handle and blade.

There are also offset screwdrivers and special screwdriver bits that attach to a ratchet or extension.

The important thing to remember when buying screwdrivers is that they really do come in sizes designed to fit different size fasteners. The slot in any screw has definite dimensions - length, width and depth. Like a bolt head or a nut, the screw slot must be driven by a tool that uses all of the available bearing surface and doesn't slip. Don't use a big wide blade on a small screw and don't try to turn a large screw slot with a tiny, narrow blade. The same principles apply to Allen heads, Phillips heads, Torx heads, etc. Don't even think of using a slotted screwdriver on one of these heads! And don't use your screwdrivers as levers, chisels or punches! This kind of abuse turns them into very bad screwdrivers.

Standard screwdrivers

These are used to remove and install conventional slotted screws and are available in a wide range of sizes denoting the width of the tip and the length of the shank (for example: a 3/8 x 10-inch screwdriver is 3/8-inch wide at the tip and the shank is 10-inches long). You should have a variety of screwdrivers so screws of various sizes can be dealt with without damaging them. The blade end must be the same width and thickness as the screw slot to work properly, without slipping. When selecting standard screwdrivers, choose good-quality tools, preferably with chrome moly, forged steel shanks. The tip of the shank should be ground to a parallel, flat profile (hollow ground) and not to a taper or wedge shape, which will tend to twist out of the slot when pressure is applied **(see illustration)**.

All screwdrivers wear in use, but standard types can be reground to shape a number of times. When reshaping a tip, start by grinding the very end flat at right angles to the shank. Make sure the tip fits snugly in the slot of a screw of the appropriate size and keep the sides of the tip parallel. Remove only a small amount of metal at a time to avoid overheating the tip and destroying the temper of the steel.

Phillips screwdrivers

Phillips screws are sometimes installed during initial assembly with air tools and are next to impossible to

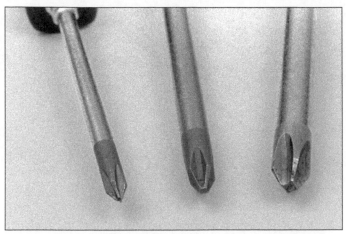

2.27 The tip size on a Phillips screwdriver is indicated by a number from 1 to 4, with 1 the smallest (left - No. 1; center - No. 2; right - No. 3)

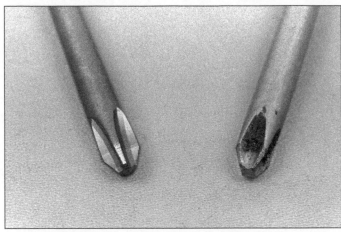

2.28 New (left) and worn (right) Phillips screwdriver tips

remove later without ruining the heads, particularly if the wrong size screwdriver is used. And don't use other types of cross-head screwdrivers (Torx, Posi-drive, etc.) on Phillips screws - they won't work.

The only way to ensure the screwdrivers you buy will fit properly, is to take a couple of screws with you to make sure the fit between the screwdriver and fastener is snug. If the fit is good, you should be able to angle the blade down almost vertically without the screw slipping off the tip. Use only screwdrivers that fit exactly - anything else is guaranteed to chew out the screw head instantly.

The idea behind all cross-head screw designs is to make the screw and screwdriver blade self-aligning. Provided you aim the blade at the center of the screw head, it'll engage correctly, unlike conventional slotted screws, which need careful alignment. This makes the screws suitable for machine installation on an assembly line (which explains why they're sometimes so tight and difficult to remove). The drawback with these screws is the driving tangs on the screwdriver tip are very small and must fit very precisely in the screw head. If this isn't the case, the huge loads imposed on small flats of the screw slot simply tear the metal away, at which point the screw ceases to be removable by normal methods. The problem is made worse by the normally soft material chosen for screws.

To deal with these screws on a regular basis, you'll need high-quality screwdrivers with various size tips so you'll be sure to have the right one when you need it. Phillips screwdrivers are sized by the tip number and length of the shank (for example: a number 2 x 6-inch Phillips screwdriver has a number 2 tip - to fit screws of only that size recess - and the shank is 6-inches long). Tip sizes 1, 2 and 3 should be adequate for engine repair work **(see illustration)**. If the tips get worn or damaged, buy new screwdrivers so the tools don't destroy the screws they're used on **(see illustration)**.

Here's a tip that may come in handy when using Phillips screwdrivers - if the screw is extremely tight and the tip tends to back out of the recess rather than turn the screw, apply a small amount of valve lapping compound to the screwdriver tip so it will grip the screw better.

Hammers

Resorting to a hammer should always be the last resort. When nothing else will do the job, a medium-size ball peen hammer, a heavy rubber mallet and a heavy soft-brass hammer **(see illustration)** are often the only way to loosen or install a part.

A ball-peen hammer has a head with a conventional cylindrical face at one end and a rounded ball end at the other and is a general-purpose tool found in almost any type of shop. It has a shorter neck than a claw hammer and the face is tempered for striking punches and chisels. A fairly large hammer is preferable to a small one. Although it's possible to find small ones, you won't need them very often and it's much easier to control the blows from a heavier head. As a general rule, a single 12 or 16-ounce hammer will work for most jobs, though occasionally larger or smaller ones may be useful.

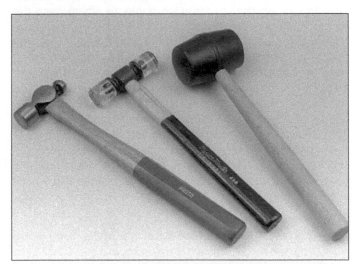

2.29 A ball-peen hammer, soft-face hammer and rubber mallet (left-to-right) will be needed for various tasks (any steel hammer can be used in place of the ball-peen hammer)

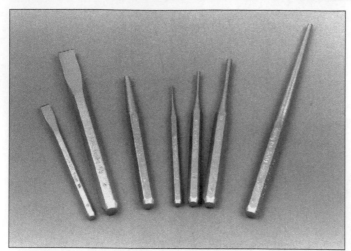

2.30 Cold chisels, center-punches, pin punches and line-up punches (left-to-right) will be needed sooner or later for many jobs

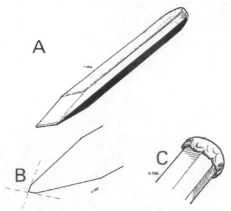

2.31 A typical general purpose cold chisel - note the angle of the cutting edge (A), which should be checked and resharpened on a regular basis; the mushroomed head (B) is dangerous and should be filed to restore it to its original shape

A soft-face hammer is used where a steel hammer could cause damage to the component or other tools being used. A steel hammer head might crack an aluminum part, but a rubber or plastic hammer can be used with more confidence. Soft-face hammers are available with interchangeable heads (usually one made of rubber and another made of relatively hard plastic). When the heads are worn out, new ones can be installed. If finances are really limited, you can get by without a soft-face hammer by placing a small hardwood block between the component and a steel hammer head to prevent damage.

Hammers should be used with common sense; the head should strike the desired object squarely and with the right amount of force. For many jobs, little effort is needed - simply allow the weight of the head to do the work, using the length of the swing to control the amount of force applied. With practice, a hammer can be used with surprising finesse, but it'll take a while to achieve. Initial mistakes include striking the object at an angle, in which case the hammer head may glance off to one side, or hitting the edge of the object. Either one can result in damage to the part or to your thumb, if it gets in the way, so be careful. Hold the hammer handle near the end, not near the head, and grip it firmly but not too tightly.

Check the condition of your hammers on a regular basis. The danger of a loose head coming off is self-evident, but check the head for chips and cracks too. If damage is noted, buy a new hammer - the head may chip in use and the resulting fragments can be extremely dangerous. It goes without saying that eye protection is essential whenever a hammer is used.

Punches and chisels

Punches and chisels (see illustration) are used along with a hammer for various purposes in the shop. Drift punches are often simply a length of round steel bar used to drive a component out of a bore in the engine or equipment it's mounted on. A typical use would be for removing or installing a bearing or bushing. A drift of the same diameter as the bearing outer race is placed against the bearing and tapped with a hammer to knock it in or out of the bore.

Most manufacturers offer special drifts for the various bearings in a particular engine. While they're useful to a busy dealer service department, they are prohibitively expensive for the do-it-yourselfer who may only need to use them once. In such cases, it's better to improvise. For bearing removal and installation, it's usually possible to use a socket of the appropriate diameter to tap the bearing in or out; an unorthodox use for a socket, but it works.

Smaller diameter drift punches can be purchased or fabricated from steel bar stock. In some cases, you'll need to drive out items like corroded engine mounting bolts. Here, it's essential to avoid damaging the threaded end of the bolt, so the drift must be a softer material than the bolt.

Brass or copper is the usual choice for such jobs; the drift may be damaged in use, but the thread will be protected.

Punches are available in various shapes and sizes and a set of assorted types will be very useful. One of the most basic is the center punch, a small cylindrical punch with the end ground to a point. It'll be needed whenever a hole is drilled. The center of the hole is located first and the punch is used to make a small indentation at the intended point. The indentation acts as a guide for the drill bit so the hole ends up in the right place. Without a punch mark the drill bit will wander and you'll find it impossible to drill with any real accuracy. You can also buy automatic center punches. They're spring loaded and are pressed against the surface to be marked, without the need to use a hammer.

Pin punches are intended for removing items like roll pins (semi-hard, hollow pins that fit tightly in their holes). Pin punches have other uses, however. You may occasionally have to remove rivets or bolts by cutting off the heads and driving out the shanks with a pin punch. They're also very handy for aligning holes in components while bolts or screws are inserted.

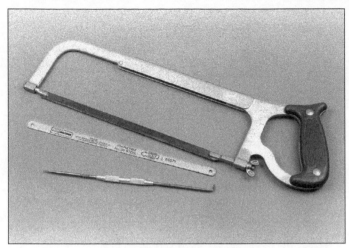

2.32 Hacksaws are handy for little cutting jobs like sheet metal and rusted fasteners

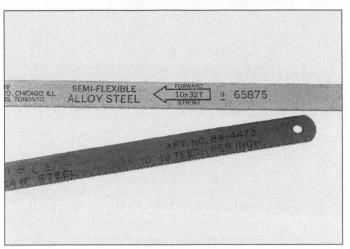

2.33 Hacksaw blades are marked with the number of teeth per inch (TPI) - use a relatively coarse blade for aluminum and thicker items such as bolts or bar stock; use a finer blade for materials like thin sheet steel

Of the various sizes and types of metal-cutting chisels available, a simple cold chisel is essential in any mechanic's workshop. One about 6-inches long with a 1/2-inch wide blade should be adequate. The cutting edge is ground to about 80-degrees **(see illustration)**, while the rest of the tip is ground to a shallower angle away from the edge. The primary use of the cold chisel is rough metal cutting - this can be anything from sheet metal work (uncommon on engines) to cutting off the heads of seized or rusted bolts or splitting nuts. A cold chisel is also useful for turning out screws or bolts with messed-up heads.

All of the tools described in this section should be good quality items. They're not particularly expensive, so it's not really worth trying to save money on them. More significantly, there's a risk that with cheap tools, fragments may break off in use - a potentially dangerous situation.

Even with good-quality tools, the heads and working ends will inevitably get worn or damaged, so it's a good idea to maintain all such tools on a regular basis. Using a file or bench grinder, remove all burrs and mushroomed edges from around the head. This is an important task because the build-up of material around the head can fly off when it's struck with a hammer and is potentially dangerous. Make sure the tool retains its original profile at the working end, again, filing or grinding off all burrs. In the case of cold chisels, the cutting edge will usually have to be reground quite often because the material in the tool isn't usually much harder than materials typically being cut. Make sure the edge is reasonably sharp, but don't make the tip angle greater than it was originally; it'll just wear down faster if you do.

The techniques for using these tools vary according to the job to be done and are best learned by experience. The one common denominator is the fact they're all normally struck with a hammer. It follows that eye protection should be worn. Always make sure the working end of the tool is in contact with the part being punched or cut. If it isn't, the tool will bounce off the surface and damage may result.

Hacksaws

A hacksaw **(see illustration)** consists of a handle and frame supporting a flexible steel blade under tension. Blades are available in various lengths and most hacksaws can be adjusted to accommodate the different sizes. The most common blade length is 10-inches.

Most hacksaw frames are adequate. There's little difference between brands. Pick one that's rigid and allows easy blade changing and repositioning.

The type of blade to use, indicated by the number of teeth per inch, (TPI) **(see illustration)**, is determined by the material being cut. The rule of thumb is to make sure at least three teeth are in contact with the metal being cut at any one time **(see illustration)**. In practice, this means a fine blade for cutting thin sheet materials, while a coarser

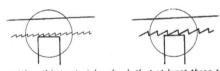

When cutting thin materials, check that at least three teeth are in contact with the workpiece at any time. Too coarse a blade will result in a poor cut and may break the blade. If you do not have the correct blade, cut at a shallow angle to the material

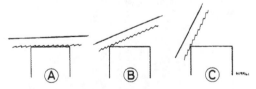

The correct cutting angle is important. If it is too shallow (A) the blade will wander. The angle shown at (B) is correct when starting the cut, and may be reduced slightly once under way. In (C) the angle is too steep and the blade will be inclined to jump out of the cut

2.34 Correct procedure for use of a hacksaw

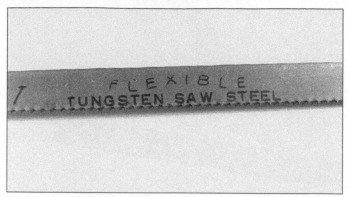

2.35 Good quality hacksaw blades are marked like this

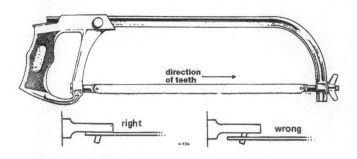

2.36 Correct installation of a hacksaw blade - the teeth must point away from the handle and butt against the locating lugs

blade can be used for faster cutting through thicker items such as bolts or bar stock. When cutting thin materials, angle the saw so the blade cuts at a shallow angle. More teeth are in contact and there's less chance of the blade binding and breaking, or teeth breaking.

When you buy blades, choose a reputable brand. Cheap, unbranded blades may be perfectly acceptable, but you can't tell by looking at them. Poor quality blades will be insufficiently hardened on the teeth edge and will dull quickly. Most reputable brands will be marked "Flexible High Speed Steel" or a similar term, to indicate the type of material used **(see illustration)**. It is possible to buy "unbreakable" blades (only the teeth are hardened, leaving the rest of the blade less brittle).

Sometimes, a full-size hacksaw is too big to allow access to a frozen nut or bolt. On most saws, you can overcome this problem by turning the blade 90-degrees. Occasionally you may have to position the saw around an obstacle and then install the blade on the other side of it. Where space is really restricted, you may have to use a handle that clamps onto a saw blade at one end. This allows access when a hacksaw frame would not work at all and has another advantage in that you can make use of broken off

hacksaw blades instead of throwing them away. Note that because only one end of the blade is supported, and it's not held under tension, it's difficult to control and less efficient when cutting.

Before using a hacksaw, make sure the blade is suitable for the material being cut and installed correctly in the frame **(see illustration)**. Whatever it is you're cutting must be securely supported so it can't move around. The saw cuts on the forward stroke, so the teeth must point away from the handle. This might seem obvious, but it's easy to install the blade backwards by mistake and ruin the teeth on the first few strokes. Make sure the blade is tensioned adequately or it'll distort and chatter in the cut and may break. Wear safety glasses and be careful not to cut yourself on the saw blade or the sharp edge of the cut.

Files

Files **(see illustration)** come in a wide variety of sizes and types for specific jobs, but all of them are used for the same basic function of removing small amounts of metal in a controlled fashion. Files are used by mechanics mainly for deburring, marking parts, removing rust, filing the heads off rivets, restoring threads and fabricating small parts.

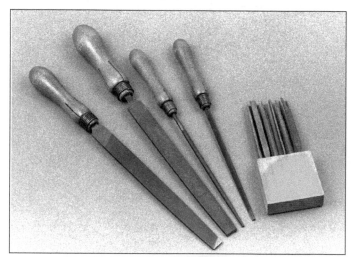

2.37 Get a good assortment of files - they're handy for deburring, marking parts, removing rust, filing the heads off rivets, restoring threads and fabricating small parts

2.38 Files are either single-cut (left) or double-cut (right) - generally speaking, use a single-cut file to produce a very smooth surface; use a double-cut file to remove large amounts of material quickly

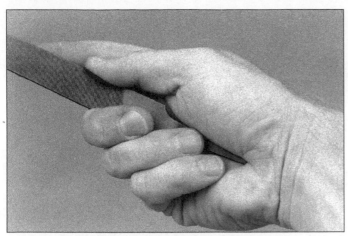

2.39 Never use a file without a handle - the tang is sharp and could puncture your hand

2.40 Adjustable handles that will work with many different size files are also available

File shapes commonly available include flat, half-round, round, square and triangular. Each shape comes in a range of sizes (lengths) and cuts ranging from rough to smooth. The file face is covered with rows of diagonal ridges which form the cutting teeth. They may be aligned in one direction only (single cut) or in two directions to form a diamond-shaped pattern (double-cut) **(see illustration)**. The spacing of the teeth determines the file coarseness, again, ranging from rough to smooth in five basic grades: Rough, coarse, bastard, second-cut and smooth.

You'll want to build up a set of files by purchasing tools of the required shape and cut as they're needed. A good starting point would be flat, half-round, round and triangular files (at least one each - bastard or second-cut types). In addition, you'll have to buy one or more file handles (files are usually sold without handles, which are purchased separately and pushed over the tapered tang of the file when in use) **(see illustration)**. You may need to buy more than one size handle to fit the various files in your tool box, but don't attempt to get by without them. A file tang is fairly sharp and you almost certainly will end up stabbing yourself in the palm of the hand if you use a file without a handle and it catches in the workpiece during use. Adjustable handles are also available for use with files of various sizes, eliminating the need for several handles **(see illustration)**.

Exceptions to the need for a handle are fine swiss pattern files, which have a rounded handle instead of a tang. These small files are usually sold in sets with a number of different shapes. Originally intended for very fine work, they can be very useful for use in inaccessible areas. Swiss files are normally the best choice if piston ring ends require filing to obtain the correct end gap.

The correct procedure for using files is fairly easy to master. As with a hacksaw, the work should be clamped securely in a vise, if needed, to prevent it from moving around while being worked on. Hold the file by the handle, using your free hand at the file end to guide it and keep it flat in relation to the surface being filed. Use smooth cutting strokes and be careful not to rock the file as it passes over the surface. Also, don't slide it diagonally across the sur-

face or the teeth will make grooves in the workpiece. Don't drag a file back across the workpiece at the end of the stroke - lift it slightly and pull it back to prevent damage to the teeth.

Files don't require maintenance in the usual sense, but they should be kept clean and free of metal filings. Steel is a reasonably easy material to work with, but softer metals like aluminum tend to clog the file teeth very quickly, which will result in scratches in the workpiece. This can be avoided by rubbing the file face with chalk before using it. General cleaning is carried out with a file card or a fine wire brush. If kept clean, files will last a long time - when they do eventually dull, they must be replaced; there is no satisfactory way of sharpening a worn file.

Taps and dies
Taps

Tap and die sets **(see illustration)** are available in inch and metric sizes. Taps are used to cut internal threads and

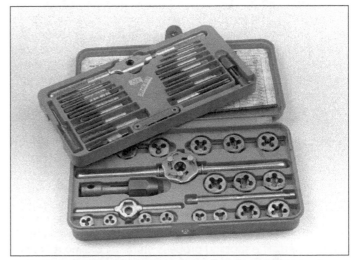

2.41 Tap and dies sets are available in inch and metric sizes - taps are used for cutting internal threads and cleaning and restoring damaged threads; dies are used for cutting, cleaning and restoring external threads

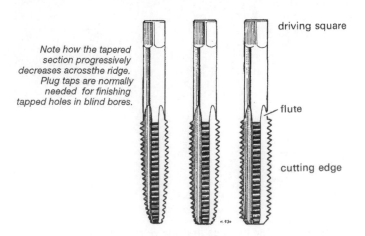

Note how the tapered section progressively decreases across the ridge. Plug taps are normally needed for finishing tapped holes in blind bores.

driving square

flute

cutting edge

2.42 Taper, plug and bottoming taps (left-to-right)

Note how the tapered section progressively decreases across the ridge. Plug taps are normally needed for finishing tapped holes in blind bores.

clean or restore damaged threads. A tap consists of a fluted shank with a drive square at one end. It's threaded along part of its length - the cutting edges are formed where the flutes intersect the threads **(see illustration)**. Taps are made from hardened steel so they will cut threads in materials softer than what they're made of.

Taps come in three different types: Taper, plug and bottoming. The only real difference is the length of the chamfer on the cutting end of the tap. Taper taps are chamfered for the first 6 or 8 threads, which makes them easy to start but prevents them from cutting threads close to the bottom of a hole. Plug taps are chamfered up about 3 to 5 threads, which makes them a good all around tap because they're relatively easy to start and will cut nearly to the bottom of a hole. Bottoming taps, as the name implies, have a very short chamfer (1-1/2 to 3 threads) and will cut as close to the bottom of a blind hole as practical. However, to do this, the threads should be started with a plug or taper tap.

Although cheap tap and die sets are available, the quality is usually very low and they can actually do more harm than good when used on threaded holes in aluminum engines. The alternative is to buy high-quality taps if and when you need them, even though they aren't cheap, especially if you need to buy two or more thread pitches in a given size. Despite this, it's the best option - you'll probably only need taps on rare occasions, so a full set isn't absolutely necessary.

Taps are normally used by hand (they can be used in machine tools, but not when doing engine repairs). The square drive end of the tap is held in a tap wrench (an adjustable T-handle). For smaller sizes, a T-handled chuck can be used. The tapping process starts by drilling a hole of the correct diameter. For each tap size, there's a corresponding twist drill that will produce a hole of the correct size. This is important; too large a hole will leave the finished thread with the tops missing, producing a weak and

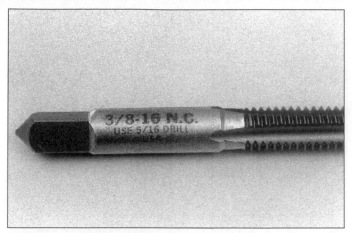

2.43 If you need to drill and tap a hole, the drill bit size to use for a given bolt (tap) is marked on the tap

unreliable grip. Conversely, too small a hole will place excessive loads on the hard and brittle shank of the tap, which can break it off in the hole. Removing a broken off tap from a hole is no fun! The correct tap drill size is normally marked on the tap itself or the container it comes in **(see illustration)**.

Dies

Dies are used to cut, clean or restore external threads. Most dies are made from a hex-shaped or cylindrical piece of hardened steel with a threaded hole in the center. The threaded hole is overlapped by three or four cutouts, which equate to the flutes on taps and allow metal waste to escape during the threading process. Dies are held in a T-handled holder (called a die stock) **(see illustration)**. Some dies are split at one point, allowing them to be adjusted slightly (opened and closed) for fine control of thread clearances.

Dies aren't needed as often as taps, for the simple reason it's normally easier to install a new bolt than to salvage one. However, it's often helpful to be able to extend the threads of a bolt or clean up damaged threads with a die. Hex-shaped dies are particularly useful for mechanic's

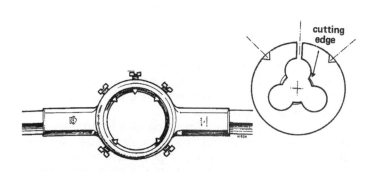

cutting edge

2.44 A die (right) is used for cutting external threads (this one is a split-type/adjustable die) and is held in a tool called a die stock (left)

2.45 Hex-shaped dies are especially handy for mechanic's work because they can be turned with a wrench

2.46 A two- or three-jaw puller will come in handy for many tasks in the shop and can also be used for working on other types of equipment

work, since they can be turned with a wrench **(see illustration)** and are usually less expensive than adjustable ones. The procedure for cutting threads with a die is broadly similar to that described above for taps. When using an adjustable die, the initial cut is made with the die fully opened, the adjustment screw being used to reduce the diameter of successive cuts until the finished size is reached. As with taps, a cutting lubricant should be used, and the die must be backed off every few turns to clear swarf from the cutouts.

Pullers

You'll need a general-purpose puller for engine rebuilding. Pullers can removed seized or corroded parts, bad bushes or bearings and dynamic balancers. Universal two- and three-legged pullers are widely available in numerous designs and sizes.

The typical puller consists of a central boss with two or three pivoting arms attached. The outer ends of the arms are hooked jaws which grab the part you want to pull off **(see illustration)**. You can reverse the arms on most pullers to use the puller on internal openings when necessary. The central boss is threaded to accept a puller bolt, which does the work. You can also get hydraulic versions of these tools which are capable of more pressure, but they're expensive.

You can adapt pullers by purchasing, or fabricating, special jaws for specific jobs. If you decide to make your own jaws, keep in mind that the pulling force should be concentrated as close to the center of the component as possible to avoid damaging it.

Before you use a puller, assemble it and check it to make sure it doesn't snag on anything and the loads on the part to be removed are distributed evenly. If you're dealing with a part held on the shaft by a nut, loosen the nut but don't remove it. Leaving the nut on helps prevent distortion of the shaft end under pressure from the puller bolt and stops the part from flying off the shaft when it comes loose.

Tighten a puller gradually until the assembly is under moderate pressure, then try to jar the component loose by striking the puller bolt a few times with a hammer. If this

doesn't work, tighten the bolt a little further and repeat the process. If this approach doesn't work, stop and reconsider. At some point you must make a decision whether to continue applying pressure in this manner. Sometimes, you can apply penetrating oil around the joint and leave it overnight, with the puller in place and tightened securely. By the next day, the taper has separated and the problem has resolved itself.

If nothing else works, try heating the area surrounding the troublesome part with a propane or gas welding torch (We don't, however, recommend messing around with welding equipment if you're not already experienced in its use). Apply the heat to the hub area of the component you wish to remove. Keep the flame moving to avoid uneven heating and the risk of distortion. Keep pressure applied with the puller and make sure that you're able to deal with the resulting hot component and the puller jaws if it does come free. Be very careful to keep the flame away from aluminum parts.

If all reasonable attempts to remove a part fail, don't be afraid to give up. It's cheaper to quit now than to repair a badly damaged engine. Either buy or borrow the correct tool, or take the engine to a dealer and ask him to remove the part for you.

Drawbolt extractors

The simple drawbolt extractor is easy to make up and invaluable in every workshop. There are no commercially available tools of this type; you simply make a tool to suit a particular application. You can use a drawbolt extractor to pull out stubborn piston pins and to remove bearings and bushings.

To make a drawbolt extractor, you'll need an assortment of threaded rods in various sizes (available at hardware stores), and nuts to fit them. You'll also need assorted washers, spacers and tubing. For things like piston pins, you'll usually need a longer piece of tube.

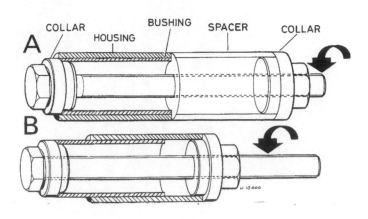

2.47 Typical drawbolt uses - in A, the nut is tightened to pull the collar and bushing into the large spacer; in B, the spacer is left out and the drawbolt is repositioned to install the new bushing

2.49 A bench vise is one of the most useful pieces of equipment you can have in the shop - bigger is usually better with vises, so get a vise with jaws that open at least four inches

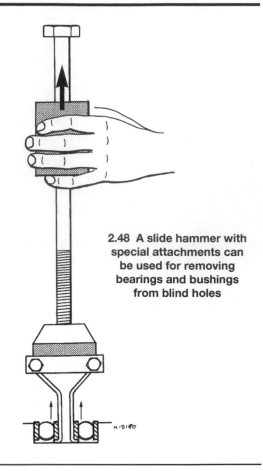

2.48 A slide hammer with special attachments can be used for removing bearings and bushings from blind holes

Some typical drawbolt uses are shown in the accompanying line drawings **(see illustration)**. They also reveal the order of assembly of the various pieces. The same arrangement, minus the tubular spacer section, can usually be used to install a new bushing or piston pin. Using the tool is quite simple. Just make sure you get the bush or pin square to the bore when you install it. Lubricate the part being pressed into place, where appropriate.

Pullers for use in blind bores

Bushings or bearings installed in "blind holes" often require special pullers. Some bearings can be removed without a puller if you heat the engine or component evenly in an oven and tap it face down on a clean wooden surface to dislodge the bearing. Wear heavy gloves to protect yourself when handling the heated components. If you need a puller to do the job, get a slide-hammer with interchange-

able tips. Slide hammers range from universal two or three-jaw puller arrangements to special bearing pullers. Bearing pullers are hardened steel tubes with a flange around the bottom edge. The tube is split at several places, which allows a wedge to expand the tool once it's in place. The tool fits inside the bearing inner race and is tightened so the flange or lip is locked under the edge of the race.

The slide-hammer consists of a steel shaft with a stop at its upper end. The shaft carries a sliding weight which slides along the shaft until it strikes the stop. This allows the tool holding the bearing to drive it out of the bore **(see illustration)**. A bearing puller set is an expensive and infrequently-used piece of equipment, so take the engine to a dealer and have the bearings/bushings replaced.

Bench vise

The bench vise **(see illustration)** is an essential tool in a shop. Buy the best quality vise you can afford. A good vise is expensive, but the quality of its materials and workmanship are worth the extra money. Size is also important - bigger vises are usually more versatile. Make sure the jaws open at least four inches. Get a set of soft jaws to fit the vise as well - you'll need them to grip engine parts that could be damaged by the hardened vise jaws **(see illustration)**.

Really, the only power tool you absolutely need is an electric drill. But if you have an air compressor and electricity, there's a wide range of pneumatic and electric hand tools to make all sorts of jobs easier and faster.

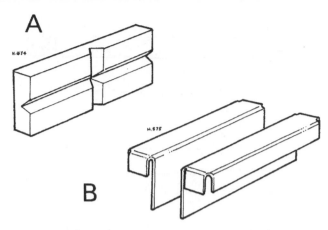

2.50 Sometimes, the parts you have to jig up in the vise are delicate, or made of soft material - to avoid damaging them, get a pair of fiberglass or plastic "soft jaws" (A) or fabricate your own with 1/8-inch thick aluminum sheet (B)

2.51 Although it's not absolutely necessary, an air compressor can make many jobs easier and produce better results, especially when air powered tools are available to use with it

Power tools

Air compressor

An air compressor **(see illustration)** makes most jobs easier and faster. Drying off parts after cleaning them with solvent, blowing out passages in a block or head, running power tools - the list is endless. Once you buy a compressor, you'll wonder how you ever got along without it. Air tools really speed up tedious procedures like removing and installing cylinder head bolts, crankshaft main bearing bolts or vibration damper (crankshaft pulley) bolts.

Bench-mounted grinder

A bench grinder **(see illustration)** is also handy. With a wire wheel on one end and a grinding wheel on the other, it's great for cleaning up fasteners, sharpening tools and removing rust. Make sure the grinder is fastened securely to the bench or stand, always wear eye protection when oper-

ating it and never grind aluminum parts on the grinding wheel.

Electric drills

Countersinking bolt holes, enlarging oil passages, honing cylinder bores, removing rusted or broken off fasteners, enlarging holes and fabricating small parts - electric drills **(see illustration)** are indispensable for engine work. A 3/8-inch chuck (drill bit holder) will handle most jobs. Collect

2.52 Another indispensable piece of equipment is the bench grinder (with a wire wheel mounted on one arbor) - make sure it's securely bolted down and never use it with the rests or eye shields removed

2.53 Electric drills can be cordless (above) or 115-volt, AC-powered (below)

2.54 Get a set of good quality drill bits for drilling holes and wire brushes of various sizes for cleaning up metal parts - make sure the bits are designed for drilling in metal

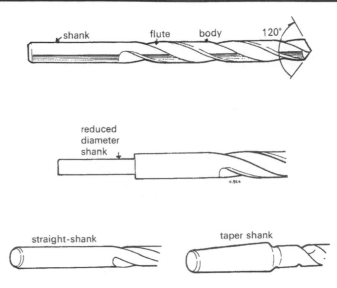

2.55 A typical drill bit (top), a reduced shank bit (center), and a tapered shank bit (bottom right)

several different wire brushes to use in the drill and make sure you have a complete set of sharp *metal* drill bits **(see illustration)**. Cordless drills are extremely versatile because they don't force you to work near an outlet. They're also handy to have around for a variety of non-mechanical jobs.

Twist drills and drilling equipment

Drilling operations are done with twist drills, either in a hand drill or a drill press. Twist drills (or drill bits, as they're often called) consist of a round shank with spiral flutes formed into the upper two-thirds to clear the waste produced while drilling, keep the drill centered in the hole and finish the sides of the hole.

The lower portion of the shank is left plain and used to hold the drill in the chuck. In this section, we will discuss only normal parallel shank drills **(see illustration)**. There is another type of bit with the plain end formed into a special size taper designed to fit directly into a corresponding socket in a heavy-duty drill press. These drills are known as

2.56 Drill bits in the range most commonly used are available in fractional sizes (left) and number sizes (right) so almost any size hole can be drilled

Morse Taper drills and are used primarily in machine shops. At the cutting end of the drill, two edges are ground to form a conical point. They're generally angled at about 60-degrees from the drill axis, but they can be reground to other angles for specific applications. For general use the standard angle is correct - this is how the drills are supplied.

When buying drills, purchase a good-quality set (sizes 1/16 to 3/8-inch). Make sure the drills are marked "High Speed Steel" or "HSS". This indicates they're hard enough to withstand continual use in metal; many cheaper, unmarked drills are suitable only for use in wood or other soft materials. Buying a set ensures the right size bit will be available when it's needed.

Twist drill sizes

Twist drills are available in a vast array of sizes, most of which you'll never need. There are three basic drill sizing systems: Fractional, number and letter **(see illustration)** (we won't get involved with the fourth system, which is metric sizes).

Fractional sizes start at 1/64-inch and increase in increments of 1/64-inch. Number drills range in descending order from 80 (0.0135-inch), the smallest, to 1 (0.2280-inch), the largest. Letter sizes start with A (0.234-inch), the smallest, and go through Z (0.413-inch), the largest.

This bewildering range of sizes means it's possible to drill an accurate hole of almost any size within reason. In practice, you'll be limited by the size of chuck on your drill (normally 3/8 or 1/2-inch). In addition, very few stores stock the entire range of possible sizes, so you'll have to shop around for the nearest available size to the one you require.

Sharpening twist drills

Like any tool with a cutting edge, twist drills will eventually get dull **(see illustration)**. How often they'll need sharpening depends to some extent on whether they're used

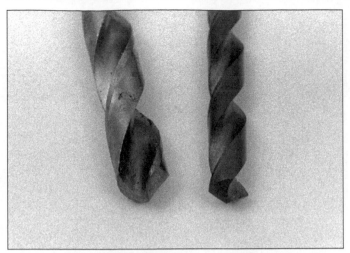

2.57 If a bit gets dull (left), discard it or resharpen it so it looks like the bit on the right

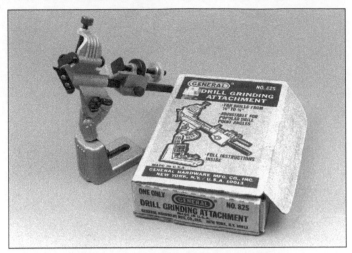

2.58 Inexpensive drill bit sharpening jigs designed to be used with a bench grinder are widely available - even if you only use it to resharpen drill bits, it'll pay for itself

correctly. A dull twist drill will soon make itself known. A good indication of the condition of the cutting edges is to watch the waste emerging from the hole being drilled. If the tip is in good condition, two even spirals of waste metal will be produced; if this fails to happen or the tip gets hot, it's safe to assume that sharpening is required.

With smaller size drills - under about 1/8-inch - it's easier and more economical to throw the worn drill away and buy another one. With larger (more expensive) sizes, sharpening is a better bet. When sharpening twist drills, the included angle of the cutting edge must be maintained at the original 120-degrees and the small chisel edge at the tip must be retained. With some practice, sharpening can be done freehand on a bench grinder, but it should be noted that it's very easy to make mistakes. For most home mechanics, a sharpening jig that mounts next to the grinding wheel should be used so the drill is clamped at the correct angle **(see illustration)**.

Drilling equipment

Tools to hold and turn drill bits range from simple, inexpensive hand-operated or electric drills to sophisticated and expensive drill presses. Ideally, all drilling should be done on a drill press with the workpiece clamped solidly in a vise. These machines are expensive and take up a lot of bench or floor space, so they're out of the question for many do-it-yourselfers. An additional problem is the fact that many of the drilling jobs you end up doing will be on the engine itself or the equipment it's mounted on, in which case the tool has to be taken to the work.

The best tool for the home shop is an electric drill with a 3/8-inch chuck. Both cordless and AC drills (that run off household current) are available. If you're purchasing one for the first time, look for a well-known, reputable brand name and variable speed as minimum requirements. A 1/4-inch chuck, single-speed drill will work, but it's worth paying a little more for the larger, variable speed type.

All drills require a key to lock the bit in the chuck. When removing or installing a bit, make sure the cord is unplugged to avoid accidents. Initially, tighten the chuck by hand, checking to see if the bit is centered correctly. This is especially important when using small drill bits which can get caught between the jaws. Once the chuck is hand tight, use the key to tighten it securely - remember to remove the key afterwards!

Preparation for drilling

If possible, make sure the part you intend to drill in is securely clamped in a vise. If it's impossible to get the work to a vise, make sure it's stable and secure. Twist drills often dig in during drilling - this can be dangerous, particularly if the work suddenly starts spinning on the end of the drill. Obviously, there's little chance of a complete engine or piece of equipment doing this, but you should make sure it's supported securely.

Start by locating the center of the hole you're drilling. Use a center punch to make an indentation for the drill bit so it won't wander. If you're drilling out a broken-off bolt, be sure to position the punch in the exact center of the bolt **(see illustration)**.

2.59 Before you drill a hole, use a centerpunch to make an indentation for the drill bit so it won't wander

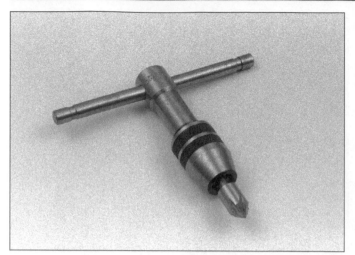

2.60 Use a large drill bit or a countersink mounted in a tap wrench to remove burrs from a hole after drilling or enlarging it

2.61 A good die grinder will deburr blocks, radius piston domes, chamfer oil holes and do a lot of other little jobs that would be tedious if done manually

If you're drilling a large hole (above 1/4-inch), you may want to make a pilot hole first. As the name suggests, it will guide the larger drill bit and minimize drill bit wandering. Before actually drilling a hole, make sure the area immediately behind the bit is clear of anything you don't want drilled.

Drilling

When drilling steel, especially with smaller bits, no lubrication is needed. If a large bit is involved, oil can be used to ensure a clean cut and prevent overheating of the drill tip. When drilling aluminum, which tends to smear around the cutting edges and clog the drill bit flutes, use kerosene as a lubricant.

Wear safety goggles or a face shield and assume a comfortable, stable stance so you can control the pressure on the drill easily. Position the drill tip in the punch mark and make sure, if you're drilling by hand, the bit is perpendicular to the surface of the workpiece. Start drilling without applying much pressure until you're sure the hole is positioned correctly. If the hole starts off center, it can be very difficult to correct. You can try angling the bit slightly so the hole center moves in the opposite direction, but this must be done before the flutes of the bit have entered the hole. It's at the starting point that a variable-speed drill is invaluable; the low speed allows fine adjustments to be made before it's too late. Continue drilling until the desired hole depth is reached or until the drill tip emerges at the other side of the workpiece.

Cutting speed and pressure are important - as a general rule, the larger the diameter of the drill bit, the slower the drilling speed should be. With a single-speed drill, there's little that can be done to control it, but two-speed or variable speed drills can be controlled. If the drilling speed is too high, the cutting edges of the bit will tend to overheat and dull. Pressure should be varied during drilling. Start with light pressure until the drill tip has located properly in the work. Gradually increase pressure so the bit cuts

evenly. If the tip is sharp and the pressure correct, two distinct spirals of metal will emerge from the bit flutes. If the pressure is too light, the bit won't cut properly, while excessive pressure will overheat the tip.

Decrease pressure as the bit breaks through the workpiece. If this isn't done, the bit may jam in the hole; if you're using a hand-held drill, it could be jerked out of your hands, especially when using larger size bits.

Once a pilot hole has been made, install the larger bit in the chuck and enlarge the hole. The second bit will follow the pilot hole - there's no need to attempt to guide it (if you do, the bit may break off). It is important, however, to hold the drill at the correct angle.

After the hole has been drilled to the correct size, remove the burrs left around the edges of the hole. This can be done with a small round file, or by chamfering the opening with a larger bit or a countersink (see illustration). Use a drill bit that's several sizes larger than the hole and simply twist it around each opening by hand until any rough edges are removed.

Enlarging and reshaping holes

The biggest practical size for bits used in a hand drill is about 1/2-inch. This is partly determined by the capacity of the chuck (although it's possible to buy larger drills with stepped shanks). The real limit is the difficulty of controlling large bits by hand; drills over 1/2-inch tend to be too much to handle in anything other than a drill press. If you have to make a larger hole, or if a shape other than round is involved, different techniques are required.

If a hole simply must be enlarged slightly, a round file is probably the best tool to use. If the hole must be very large, a hole saw will be needed, but they can only be used in sheet metal.

Large or irregular-shaped holes can also be made in sheet metal and other thin materials by drilling a series of small holes very close together. In this case the desired hole size and shape must be marked with a scribe. The next

2.62 Buy at least one fire extinguisher before you open shop - make sure it's rated for flammable liquid fires and KNOW HOW TO USE IT!

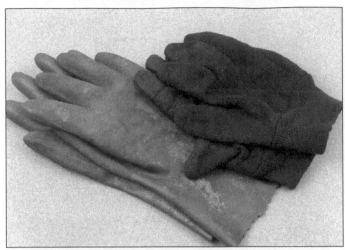

2.63 Get a pair of heavy work gloves for handling hot or sharp-edged objects and a pair of rubber gloves for washing parts with solvent

2.64 One of the most important items you'll need in the shop is a face shield or safety goggles, especially when you're hitting metal parts with a hammer, washing parts in solvent or grinding something on the bench grinder

step depends on the size bit to be used; the idea is to drill a series of almost touching holes just inside the outline of the large hole. Center punch each location, then drill the small holes. A cold chisel can then be used to knock out the waste material at the center of the hole, which can then be filed to size. This is a time consuming process, but it's the only practical approach for the home shop. Success is dependent on accuracy when marking the hole shape and using the center punch.

High-speed grinders

A good die grinder (see illustration) will deburr blocks, radius piston domes and chamfer oil holes ten times as fast as you can do any of these jobs by hand.

Safety items that should be in every shop

Fire extinguishers

Buy at least one fire extinguisher (see illustration) before doing any maintenance or repair procedures. Make sure it's rated for flammable liquid fires. Familiarize yourself with its use as soon as you buy it - don't wait until you need it to figure out how to use it. And be sure to have it checked and recharged at regular intervals. Refer to the safety tips at the end of this chapter for more information about the hazards of gasoline and other flammable liquids.

Gloves

If you're handling hot parts or metal parts with sharp edges, wear a pair of industrial work gloves to protect yourself from burns, cuts and splinters (see illustration). Wear a

pair of heavy duty rubber gloves (to protect your hands when you wash parts in solvent.

Safety glasses or goggles

Never work on a bench or high-speed grinder without safety glasses (see illustration). Don't take a chance on getting a metal sliver in your eye. It's also a good idea to wear safety glasses when you're washing parts in solvent.

Special diagnostic tools

These tools do special diagnostic tasks. They're indispensable for determining the condition of your engine. If you don't think you'll use them frequently enough to justify the investment, try to borrow or rent them. Or split the cost with a friend who also wants to get into engine rebuilding. We've listed only those tools and instruments generally available to the public, not the special tools manufactured by Chevrolet for its dealer service departments. Occasional

2.65 The combustion leak block tester tests for cracked blocks, leaky gaskets, cracked heads and warped heads by detecting combustion gases in the coolant

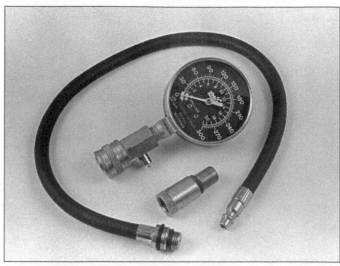

2.66 The compression gauge indicates cylinder pressure in the combustion chamber - get the kind that screws into the spark plug hole, not the type that you must push into the spark plug hole and hold in place against cylinder pressure

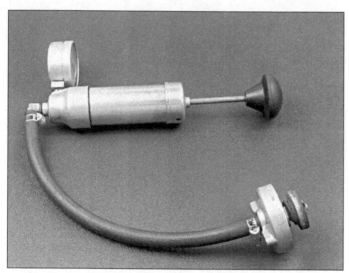

2.67 The cooling system pressure tester checks for leaks in the cooling system by pressurizing the system and measuring the rate at which it leaks down

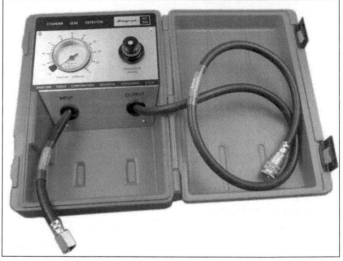

2.68 The leakdown tester indicates the rate at which pressure leaks past the piston rings, the valves and/or the head gasket

references to Chevrolet special tools may be included in the text of this manual. But we'll try to provide you with another way of doing the job without the special tool, if possible. However, when there's no alternative, you'll have to borrow or buy the tool or have the job done by a professional.

Combustion leak block tester

The combustion leak block tester (see illustration) tests for cracked blocks, leaky gaskets, cracked heads and warped heads by detecting combustion gasses in the coolant. For information on using this tool, see Chapter 3.

Compression gauge

The compression gauge (see illustration) indicates

cylinder pressure in the combustion chamber. Get the kind that screws into the spark plug hole - not the type with the universal rubber tip that you push into the plug hole. It's hard to prevent cylinder pressure from leaking past the rubber tip, especially when it's worn.

Cooling system pressure tester

The cooling system pressure tester (see illustration) checks for leaks in the cooling system.

Leakdown tester

The leakdown tester (see illustration) indicates the rate at which pressure leaks past the piston rings, the valves and/or the head gasket, in a combustion chamber.

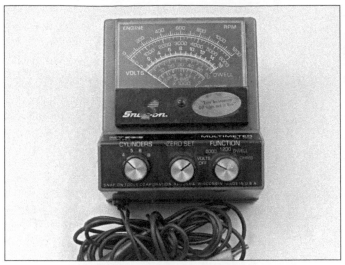

2.69 The multimeter combines the functions of a volt-meter, ammeter and ohmmeter into one unit - it can measure voltage, amperage or resistance in an electrical circuit

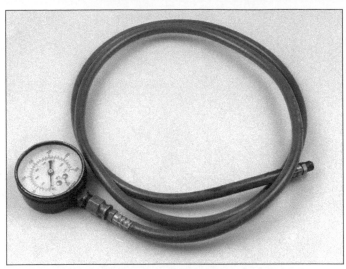

2.70 The oil pressure gauge, which screws into an oil pressure sending unit hole in the block or head, measures the oil pressure in the lubrication system

Multimeter

The multimeter **(see illustration)** combines the functions of a voltmeter, ammeter and ohmmeter into one unit. It can measure voltage, amperage or resistance in an electrical circuit.

Oil pressure gauge

The oil pressure gauge **(see illustration)** indicates the oil pressure in the lubrication system.

Spark tester

Sometimes you just want to know whether you've got spark. This handy little device **(see illustration)** looks like a spark plug and plugs into a spark plug boot just like a regu-

lar spark plug. It's got a clip which you can clip onto the nearest good ground. Just start the engine and see if the tester emits a spark across the gap between its electrode and the spark tester body.

Stethoscope

The stethoscope **(see illustration)** amplifies engine sounds, allowing you to pinpoint possible sources of pending trouble such as bad bearings, excessive play in the crank, rod knock, etc.

Tachometer/dwell meter

The tach indicates the speed at which the engine crankshaft is turning, in revolutions per minute (rpm); the

2.71 The spark tester, which looks like a spark plug, can be plugged into each plug boot and clipped to a good ground on the block or head to tell you if each spark plug wire is sending spark to its respective cylinder

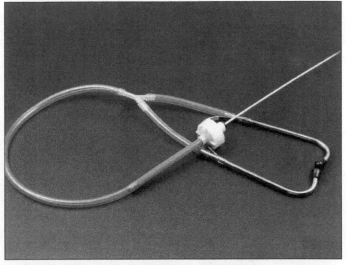

2.72 The stethoscope amplifies engine sounds, allowing you to pinpoint possible sources of trouble such as bad bearings, excessive play in the crank, rod knocks, etc.

2.73 The tach/dwell meter combines the functions of a tachometer and dwell meter into one package - the tach indicates the speed - in rpm - at which the engine crankshaft is turning; the dwell meter indicates the number of degrees of distributor rotation during which the breaker points are closed

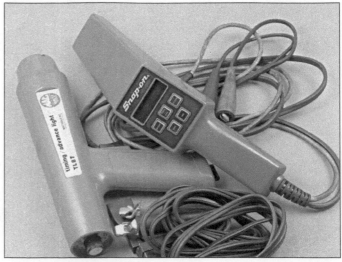

2.74 The timing light helps you synchronize ignition timing with engine timing - get a stroboscopic unit with an inductive pick-up that clamps over the No. 1 spark plug wire

dwell function indicates the number of degrees of distributor rotation during which the breaker points are closed **(see illustration)**.

Timing light

The timing light **(see illustration)** enables you to synchronize ignition timing with engine timing. Get the stroboscopic type, with an inductive pick-up that clamps over the No. 1 spark plug wire.

Vacuum gauge

The vacuum gauge **(see illustration)** indicates intake manifold vacuum, in inches of mercury (in-Hg).

Vacuum/pressure pump

The hand-operated vacuum/pressure pump **(see illustration)** can create a vacuum, or build up pressure, in a circuit to check components that are vacuum or pressure operated.

Engine rebuilding tools

Engine hoist

Get an engine hoist **(see illustration)** that's strong enough to easily lift your engine in and out of the engine compartment. A V8 is far too heavy to remove and install any other way. And you don't even want to think about the possibility of dropping an engine!

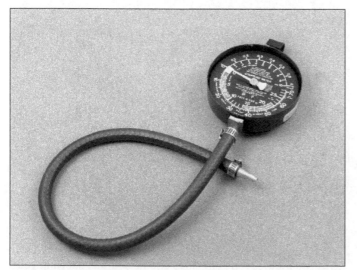

2.75 The vacuum gauge indicates intake manifold vacuum, in inches of mercury (in-Hg)

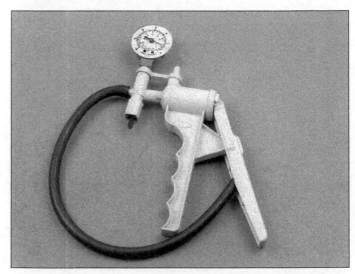

2.76 The vacuum/pressure pump can create a vacuum in a circuit, or pressurize it, to simulate the actual operating conditions

2.78 Get an engine stand sturdy enough to firmly support the engine while you're working on it - stay away from three-wheeled models - they have a tendency to tip over more easily - get a four-wheeled unit

2.77 Get an engine hoist that's strong enough to easily lift your engine in and out of the engine compartment - an adapter, like the one shown here (arrow), can be used to change the angle of the engine as it's being removed or installed

2.79 A clutch alignment tool is necessary if you plan to install a rebuilt engine mated to a manual transmission

Engine stand

A V8 is too heavy and bulky to wrestle around on the floor or the workbench while you're disassembling and reassembling it. Get an engine stand **(see illustration)** sturdy enough to firmly support the engine. Even if you plan to work on small blocks, it's a good idea to buy a stand beefy enough to handle big blocks as well. Try to buy a stand with four wheels, not three. The center of gravity of a stand is high, so it's easier to topple the stand and engine when you're cinching down head bolts with a two-foot long torque wrench. Get a stand with big casters. The larger the wheels, the easier it is to roll the stand around the shop. Wheel locks are also a good feature to have. If you want the stand out of the way between jobs, look for one that can be knocked down and slid under a workbench. Engine stands are available at rental yards; however, the cost of rental over the long period of time you'll need the stand (often weeks) is frequently as high as buying a stand.

Clutch alignment tool

The clutch alignment tool **(see illustration)** is used to center the clutch disc on the flywheel.

Cylinder surfacing hone

After boring the cylinders, you need to put a crosshatch pattern on the cylinder walls to help the new rings seat

2.80 A cylinder hone like the one shown here is easier to use, but not as versatile as the type that has three spring-loaded stones

2.81 If there are ridges at the top of the cylinder wall created by the thrust side of the piston at the top of its travel, use a ridge reamer to remove them

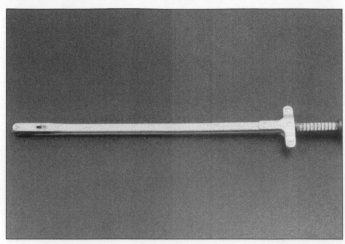

2.82 Here's a typical hydraulic lifter removal tool

2.83 The best universal ring compressor is the plier-type - to fit different bore sizes, simply insert different size compressor bands in the plier handles

2.84 The band-type ring compressor is as easy to use as the plier-type, but is more likely to snag a ring

properly. A flex hone with silicone carbide balls laminated onto the end of wire bristles (see illustration) will give you that pattern. Even if you don't bore the cylinders, you must hone the cylinders, since it breaks the glaze that coats the cylinder walls.

Cylinder ridge reamer

If the engine has a lot of miles, the top compression ring will likely wear the cylinder wall and create a ridge at the top of each cylinder (the unworn portion of the bore forms the ridge). Carbon deposits make the ridge even more pronounced. The ridge reamer (see illustration) cuts away the ridge so you can remove the piston from the top of the cylinder without damaging the ring lands.

Hydraulic lifter removal tool

Sometimes the lifters are gummed up with varnish and become stuck in their lifter bores. This tool (see illustration) will get them out.

Piston ring compressor

Trying to install the pistons without a ring compressor is almost impossible.

The best "universal" ring compressor is the plier-type, like the one manufactured by K-D Tools (see illustration). To accommodate different bore sizes, simply insert different size compressor bands in the plier handles. Plier-type spring compressors allow you to turn the piston with one hand while tapping it into the cylinder bore with a hammer handle.

The band type ring compressor (see illustration) is the cheapest of the two, and will work on a range of piston sizes, but it's more likely to snag a ring.

Piston ring groove cleaner

If you're reusing old pistons, you'll want to clean the

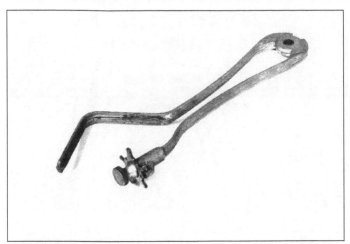

2.85 Sometimes you find that you can re-use the same pistons even when the rings need to be replaced - but the piston ring grooves in the old pistons are usually filthy - to clean them properly, you may need a piston ring groove cleaning tool

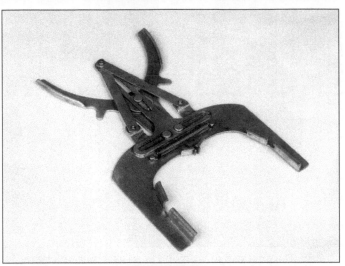

2.86 The piston ring expander pushes the ends of each ring apart so you can slip it over the piston crown and into its groove without scratching the piston or damaging the ring

carbon out of the ring grooves. This odd-looking tool **(see illustration)** has a cutting bit that digs the stuff out.

Piston ring expander

This plier-like tool **(see illustration)** pushes the ends of the ring apart so you can slip it over the piston crown and into its groove.

Valve spring compressor

The valve spring compressor **(see illustration)** compresses the valve springs so you can remove the keepers and the retainer. For engine overhaul, if you can afford it, get a C-clamp type that's designed to de-spring the head when it's off the engine. Cheaper types also work, but they're more time consuming to use.

Precision measuring tools

Think of the tools in the following list as the final stage of your tool collection. If you're planning to rebuild an engine, you've probably already accumulated all the screwdrivers, wrenches, sockets, pliers and other everyday hand tools that you need. You've also probably collected all the special-purpose tools necessary to tune and service your specific engine. Now it's time to round up the stuff you'll need to do your own measurements when you rebuild that engine.

The tool pool strategy

If you're reading this book, you may be a motorhead, but engine rebuilding isn't your life - it's an avocation. You may just want to save some money, have a little fun and learn something about engine building. If that description fits your level of involvement, think about forming a "tool pool" with a friend or neighbor who wants to get into engine rebuilding, but doesn't want to spend a lot of money. For example, you can buy a set of micrometers and the other guy can buy a dial indicator and a set of small hole gauges.

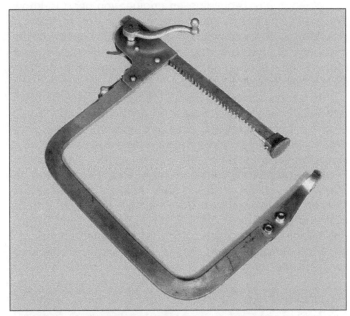

2.87 The valve spring compressor compresses the valve springs so you can remove the keepers and the retainer - the C-type (shown) reaches around to the underside of the head and pushes against the valve as it compresses the spring

Start with the basics

It would be great to own every precision measuring tool listed here, but you don't really need a machinist's chest crammed with exotic calipers and micrometers. You can often get by just fine with nothing more than a feeler gauge, modeling clay and Plastigage. Even most professional engine builders use only three tools 95-percent of the time: a one-inch outside micrometer, a dial indicator and a six-inch dial caliper. So start your collection with these three items.

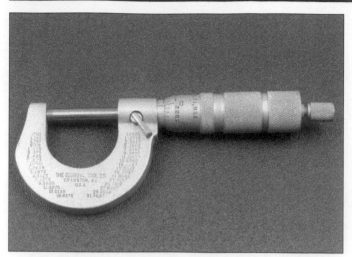

2.88 The one-inch micrometer is an essential precision measuring device for determining the dimensions of a wrist pin, valve spring shim, thrust washer, etc.

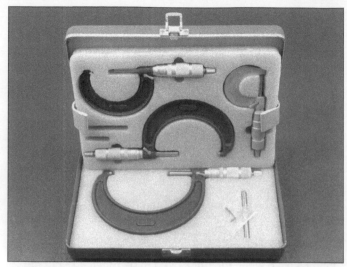

2.89 Get a good-quality micrometer set if you can afford it - this set has four micrometers ranging in size from one to four inches

Micrometers

When you're rebuilding an engine, you need to know the exact thickness of a sizeable number of pieces. Whether you're measuring the diameter of a wrist pin or the thickness of a valve spring shim or a thrust washer, your tool of choice should be the trusty one-inch outside micrometer **(see illustration)**.

Insist on accuracy to within one ten-thousandths of an inch (0.0001-inch) when you shop for a micrometer. You'll probably never need that kind of precision, but the extra decimal place will help you decide which way to round off a close measurement.

High-quality micrometers have a range of one inch. Eventually, you'll want a set **(see illustration)** that spans four, or even five, ranges: 0 to 1-inch, 1 to 2-inch, 2 to 3-inch and 3-to-4-inch. On 340 cu. in. engines and larger, you'll also probably need a 4-to-5-inch. These five micrometers will measure the thickness of any part that needs to be measured for an engine rebuild. You don't have to run

out and buy all five of these babies at once. Start with the one-inch model; then, when you have the money, get the next size you need (the 3-to 4-inch size or 4-to-5-inch is a good second choice - they measure piston diameters).

Digital micrometers **(see illustration)** are easier to read than conventional micrometers, are just as accurate and are finally starting to become affordable. If you're uncomfortable reading a conventional micrometer **(see sidebar)**, then get a digital.

Unless you're not going to use them very often, stay away from micrometers with interchangeable anvils **(see illustration)**. In theory, one of these beauties can do the work of five or six single-range micrometers. The trouble is, they're awkward to use when measuring little parts, and changing the anvils is a hassle.

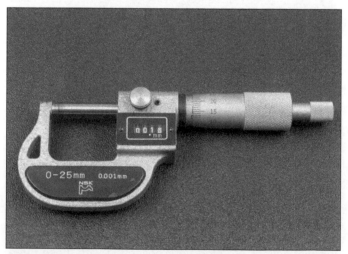

2.90 Digital micrometers are easier to read than conventional micrometers and are just as accurate

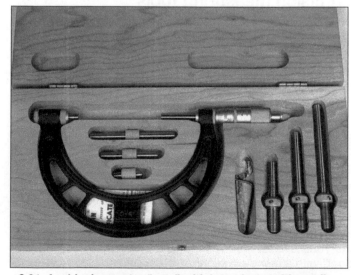

2.91 Avoid micrometer "sets" with interchangeable anvils - they're awkward to use when measuring little parts and changing the anvils is a hassle

How to read a micrometer

The outside micrometer is without a doubt the most widely used precision measuring tool. It can be used to make a variety of highly accurate measurements without much possibility of error through misreading, a problem associated with other measuring instruments, such as vernier calipers.

Like any slide caliper, the outside micrometer uses the "double contact" of its spindle and anvil **(see illustration)** touching the object to be measured to determine that object's dimensions. Unlike a caliper, however, the micrometer also features a unique precision screw adjustment which can be read with a great deal more accuracy than calipers.

Why is this screw adjustment so accurate? Because years ago toolmakers discovered that a screw with 40 precision machined threads to the inch will advance one-fortieth (0.025) of an inch with each complete turn. The screw threads on the spindle revolve inside a fixed nut concealed by a sleeve.

On a one-inch micrometer, this sleeve is engraved longitudinally with exactly 40 lines to the inch, to correspond with the number of threads on the spindle. Every fourth line is made longer and is numbered one-tenth inch, two-tenths, etc. The other lines are often staggered to make them easier to read.

The thimble (the barrel which moves up and down the sleeve as it rotates) is divided into 25 divisions around the circumference of its beveled edge and is numbered from zero to 25. Close the micrometer spindle until it touches the anvil: You should see nothing but the zero line on the sleeve next to the beveled edge of the thimble. And the zero line of the thimble should be aligned with the horizontal (or axial) line on the sleeve. Remember: Each full revolution of the spindle from zero to zero advances or retracts the spindle one-fortieth or 0.025-inch. Therefore, if you

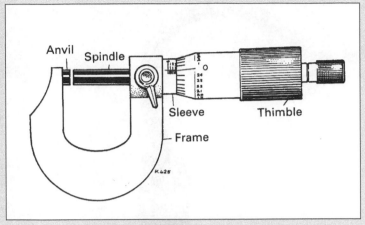

2.92 This diagram of a typical one-inch micrometer shows its major components

rotate the thimble from zero on the beveled edge to the first graduation, you will move the spindle 1/25th of 1/40th, or 1/25th of 25/1000ths, which equals 1/1000th, or 0.001-inch.

Remember: Each numbered graduation on the sleeve represents 0.1-inch, each of the other sleeve graduations represents 0.025-inch and each graduation on the thimble represents 0.001-inch. Remember those three and you're halfway there.

For example: Suppose the 4 line is visible on the sleeve. This represents 0.400-inch. Then suppose there are an additional three lines (the short ones without numbers) showing. These marks are worth 0.025-inch each, or 0.075-inch. Finally, there are also two marks on the beveled edge of the thimble beyond the zero mark, each good for 0.001-inch, or a total of 0.002-inch. Add it all up and you get 0.400 plus 0.075 plus 0.002, which equals 0.477-inch.

Some beginners use a "dollars, quarters and cents" analogy to simplify reading a micrometer. Add up the bucks and change, then put a decimal point instead of a dollar sign in front of the sum!

Dial indicators

The dial indicator **(see illustration)** is another measuring mainstay. It's indispensable for degreeing camshafts, measuring valve lift, piston deck clearances, crankshaft endplay and all kinds of other little measurements. Make sure the dial indicator you buy has a probe with at least one inch of travel, graduated in 0.001-inch increments. And get a good assortment of probe extensions up to about six inches long. Sometimes, you need to screw a bunch of these extensions together to reach into tight areas like pushrod holes.

Buy a dial indicator set that includes a flexible fixture

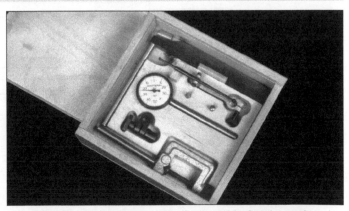

2.93 The dial indicator is indispensable for degreeing crankshafts, measuring valve lift, piston deck clearance, crankshaft endplay and a host of other critical measurements

2.94 Get an adjustable, flexible fixture like this one, and a magnetic base, to ensure maximum versatility from your dial indicator

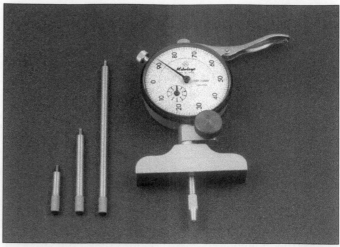

2.95 This dial indicator is designed to measure depth, such as deck height when the piston is below the block surface - with a U-shaped bridge (the base seen here is removable), you can measure the deck height of pistons that protrude above the deck (U-shaped bridges are also useful for checking the flatness of a block or cylinder head)

and a magnetic stand **(see illustration)**. If the model you buy doesn't have a magnetic base, buy one separately. Make sure the magnet is plenty strong. If a weak magnet comes loose and the dial indicator takes a tumble on a concrete floor, you can kiss it good-bye. Make sure the arm that attaches the dial indicator to the flexible fixture is sturdy and the locking clamps are easy to operate.

Some dial indicators are designed to measure depth **(see illustration)**. They have a removable base that straddles a hole. This setup is indispensable for measuring deck height when the piston is below the block surface. To measure the deck height of pistons that protrude above the deck, you'll also need a U-shaped bridge for your dial indicator. The bridge is also useful for checking the flatness of a block or a cylinder head.

Calipers

Vernier calipers **(see illustration)** aren't quite as accurate as a micrometer, but they're handy for quick measurements and they're relatively inexpensive. Most calipers have inside and outside jaws, so you can measure the inside diameter of a hole, or the outside diameter of a part.

Better-quality calipers have a dust shield over the geared rack that turns the dial to prevent small metal particles from jamming the mechanism. Make sure there's no play in the moveable jaw. To check, put a thin piece of metal between the jaws and measure its thickness with the metal close to the rack, then out near the tips of the jaws. Compare your two measurements. If they vary by more than 0.001-inch, look at another caliper - the jaw mechanism is deflecting.

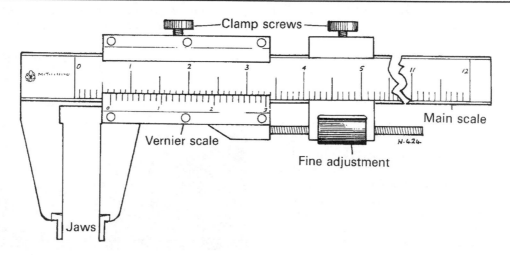

2.96 Vernier calipers aren't quite as accurate as micrometers, but they're handy for quick measurements and relatively inexpensive, and because they've got jaws that can measure internal and external dimensions, they're versatile

2.97 Dial calipers are a lot easier to read than conventional vernier calipers, particularly if your eyesight isn't as good as it used to be!

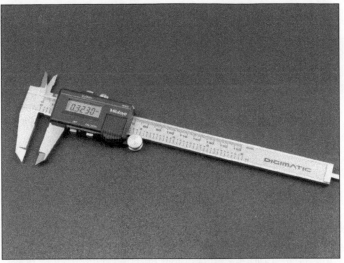

2.98 The latest calipers have a digital readout that is even easier to read than a dial caliper - another advantage of digital calipers is that they have a small microchip that allows them to convert instantaneously from inch to metric dimensions

If your eyes are going bad, or already are bad, vernier calipers can be difficult to read. Dial calipers **(see illustration)** are a better choice. Dial calipers combine the measuring capabilities of micrometers with the convenience of dial indicators. Because they're much easier to read quickly than vernier calipers, they're ideal for taking quick measurements when absolute accuracy isn't necessary. Like conventional vernier calipers, they have both inside and outside jaws which allow you to quickly determine the diameter of a hole or a part. Get a six-inch dial caliper, graduated in 0.001-inch increments.

The latest calipers **(see illustration)** have a digital LCD display that indicates both inch and metric dimensions. If you can afford one of these, it's the hot setup.

How to read a vernier caliper

On the lower half of the main beam, each inch is divided into ten numbered increments, or tenths (0.100-inch, 0.200-inch, etc.). Each tenth is divided into four increments of 0.025-inch each. The vernier scale has 25 increments, each representing a thousandth (0.001) of an inch.

First read the number of inches, then read the number of tenths. Add to this 0.025-inch; for each additional graduation. Using the English vernier scale, determine which graduation of the vernier lines up exactly with a graduation on the main beam. This vernier graduation is the number of thousandths which are to be added to the previous readings.

For example, let's say:

1) The number of inches is zero, or 0.000-inch;
2) The number of tenths is 4, or 0.400-inch;
3) The number of 0.025's is 2, or 0.050-inch; and
4) The vernier graduation which lines up with a graduation on the main beam is 15, or 0.015-inch.
5) Add them up:
   0.000
   0.400
   0.050
   0.015
6) And you get:
   0.46-inch
That's all there is to it!

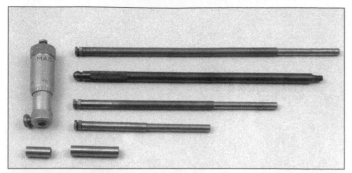

2.99 Inside micrometers are handy for measuring holes with thousandth-of-an-inch accuracy

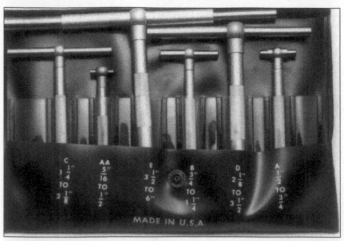

2.100 Telescoping snap gauges are used to measure smaller holes - simply insert them into a hole, turn the knurled handle to release their spring-loaded probes out to the wall, turn the handle to lock the probes into position, pull out the gauge and measure the length from the tip of one probe to the tip of the other probe with a micrometer

Inside micrometers

Cylinder bores, main bearing bores, connecting rod big ends, valve guides - automotive engines have a lot of holes that must be measured accurately within a thousandth of an inch. Inside micrometers **(see illustration)** are used for these jobs. You read an inside micrometer the same way you read an outside micrometer. But it takes more skill to get an accurate reading.

To measure the diameter of a hole accurately, you must find the widest part of the hole. This involves expanding the micrometer while rocking it from side to side and moving it up and down. Once the micrometer is adjusted properly, you should be able to pull it through the hole with a slight drag. If the micrometer feels loose or binds as you pull it through, you're not getting an accurate reading.

Fully collapsed, inside micrometers can measure holes as small as one inch in diameter. Extensions or spacers are added for measuring larger holes.

Telescoping snap gauges **(see illustration)** are used to measure smaller holes. Simply insert them into a hole and

turn the knurled handle to release their spring-loaded probes, which expand out to the walls of the hole, turn the handle the other way and lock the probes into position, then pull the gauge out. After the gauge is removed from the hole, measure its width with an outside micrometer.

For measuring really small holes, such as valve guides, you'll need a set of small hole gauges **(see illustration).** They work the same way as telescoping snap gauges, but instead of spring-loaded probes, they have expanding flanges on the end that can be screwed in and out by a threaded handle.

Dial bore gauge

The dial bore gauge **(see illustration)** is more accurate

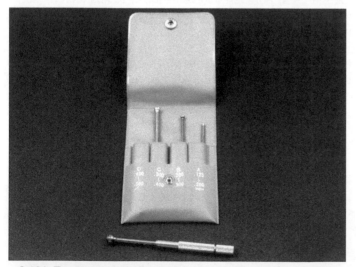

2.101 To measure really small holes, such as valve guides, you need a set of small hole gauges - to use them, simply stick them into the hole, turn the knurled handle until the expanding flanges are contacting the walls of the hole, pull out the gauge and measure the width of the gauge at the flanges with a micrometer

2.102 The dial bore gauge is more accurate and easier to use than an inside micrometer or telescoping snap gauges, but it's expensive - using various extensions, most dial gauges have a range of measurement from just over one inch to six inches or more

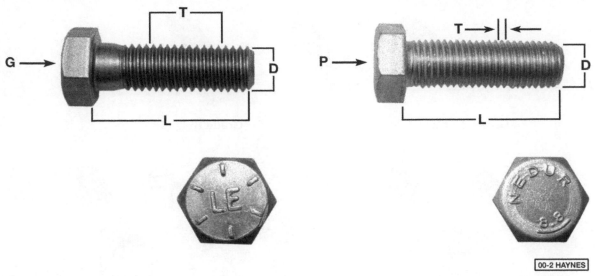

2.103 Standard (SAE and USS) bolt dimensions/grade marks

G   Grade marks (bolt strength)
L   Length (in inches)
T   Thread pitch (number of threads per inch)
D   Nominal diameter (in inches)

Metric bolt dimensions/grade marks

P   Property class (bolt strength)
L   Length (in millimeters)
T   Thread pitch (distance between threads in millimeters)
D   Diameter

and easier to use - but more expensive - than an inside micrometer for checking the roundness of the cylinders, and the bearing bores in main bearing saddles and connecting rods. Using various extensions, most dial bore gauges have a range of just over 1-inch in diameter to 6 inches or more. Unlike outside micrometers with interchangeable anvils, the accuracy of bore gauges with interchangeable extensions is unaffected. Bore gauges accurate to 0.0001-inch are available, but they're very expensive and hard to find. Most bore gauges are graduated in 0.0005-inch increments. If you use them properly, this accuracy level is more than adequate.

Storage and care of tools

Good tools are expensive, so treat them well. After you're through with your tools, wipe off any dirt, grease or metal chips and put them away. Don't leave tools Lying around in the work area. General purpose hand tools - screwdrivers, pliers, wrenches and sockets - can be hung on a wall panel or stored in a tool box. Store precision measuring instruments, gauges, meters, etc. in a tool box to protect them from dust, dirt, metal chips and humidity.

Fasteners

Fasteners - nuts, bolts, studs and screws - hold parts together. Keep the following things in mind when working with fasteners: All threaded fasteners should be clean and straight, with good threads and unrounded corners on the hex head (where the wrench fits). Make it a habit to replace all damaged nuts and bolts with new ones. Almost all fasteners have a locking device of some type, either a lock-

washer, locknut, locking tab or thread adhesive. Don't reuse special locknuts with nylon or fiber inserts. Once they're removed, they lose their locking ability. Install new locknuts.

Flat washers and lockwashers, when removed from an assembly, should always be replaced exactly as removed. Replace any damaged washers with new ones. Never use a lockwasher on any soft metal surface (such as aluminum), thin sheet metal or plastic.

Apply penetrant to rusted nuts and bolts to loosen them up and prevent breakage. Some mechanics use turpentine in a spout-type oil can, which works quite well. After applying the rust penetrant, let it work for a few minutes before trying to loosen the nut or bolt. Badly rusted fasteners may have to be chiseled or sawed off or removed with a special nut breaker, available at tool stores.

If a bolt or stud breaks off in an assembly, it can be drilled and removed with a special tool commonly available for this purpose. Most automotive machine shops can perform this task, as well as other repair procedures, such as the repair of threaded holes that have been stripped out.

Fastener Sizes

For a number of reasons, automobile manufacturers are making wider and wider use of metric fasteners. Therefore, it's important to be able to tell the difference between standard (sometimes called USS or SAE) and metric hardware, since they cannot be interchanged.

All bolts, whether standard or metric, are sized in accordance with their diameter, thread pitch and length **(see illustration)**. For example, a standard 1/2- 13 x 1 bolt is 1/2 inch in diameter, has 13 threads per inch and is 1 inch long. An M12-1.75x25 metric bolt is 12mm in diameter, has

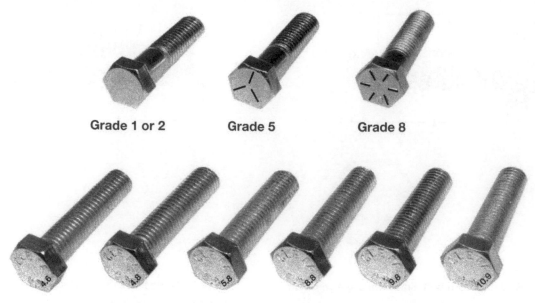

Grade 1 or 2   Grade 5   Grade 8

2.104 Bolt strength markings (top - standard/ SAE; bottom - metric)

a thread pitch of 1.75 mm (the distance between threads) and is 25 mm long. The two bolts are nearly identical, and easily confused, but they are not interchangeable.

In addition to the differences in diameter, thread pitch and length, metric and standard bolts can also be distinguished by examining the bolt heads. The distance across the flats on a standard bolt head is measured in inches; the same dimension on a metric bolt or nut is sized in millimeters. So don't use a standard wrench on a metric bolt, or vice versa.

Most standard bolts also have slashes radiating out from the center of the head **(see illustration 2.104)** to denote the grade or strength of the bolt, which is an indication of the amount of torque that can be applied to it. The greater the number of slashes, the greater the strength of the bolt. Grades 0 through 8 are commonly used on auto-

mobiles. Metric bolts have a property class (grade) number, rather than a slash, molded into their heads to indicate bolt strength. In this case, the higher the number, the stronger the bolt. Property class numbers 8.8,9.8 and 10.9 are commonly used on automobiles.

Strength markings can also be used to distinguish standard hex nuts from metric hex nuts. Many standard nuts have dots stamped into one side, while metric nuts are marked with a number **(see illustrations)**. The greater the number of dots, or the higher the number, the greater the strength of the nut.

Metric studs are also marked on their ends **(see illustration)** according to property class (grade). Larger studs are numbered (the same as metric bolts), while smaller studs carry a geometric code to denote grade.

It should be noted that many fasteners, especially

Grade	Identification
Hex Nut Grade 5	3 Dots
Hex Nut Grade 8	6 Dots

2.105a Standard hex nut strength markings

Grade	Identification
Hex Nut Property Class 9	Arabic 9
Hex Nut Property Class 10	Arabic 10

2.105b Metric nut strength markings

Class 10.9   Class 9.8   Class 8.8

00-1 HAYNES

2.106 Metric stud strength markings

Grades 0 through 2, have no distinguishing marks on them. When such is the case, the only way to determine whether it's standard or metric is to measure the thread pitch or compare it to a known fastener of the same size.

Standard fasteners are often referred to as SAE, as opposed to metric. However, it should be noted that SAE technically refers to a non-metric fine thread fastener only. Coarse thread non-metric fasteners are referred to as US sizes.

Since fasteners of the same size (both standard and metric) may have different strength ratings, be sure to reinstall any bolts, studs or nuts removed from your vehicle in their original locations. Also, when replacing a fastener with a new one, make sure that the new one has a strength rating equal to or greater than the original.

Tightening sequences and procedures

Most threaded fasteners should be tightened to a specific torque value **(see charts below)**. Torque is the twisting force applied to a threaded component such as a nut or bolt. Overtightening the fastener can weaken it and cause it to break, while undertightening can cause it to eventually come loose. Bolts, screws and studs, depending on the material they are made of and their thread diameters, have specific torque values, many of which are noted in the Specifications at the beginning of each Chapter. Be sure to follow the torque recommendations closely. For fasteners not assigned a specific torque, a general torque value chart is presented here as a guide. These torque values are for dry (unlubricated) fasteners threaded into steel or cast iron (not aluminum). As was previously mentioned, the size and grade of a fastener determine the amount of torque that can safely be applied to it. The figures listed here are approximate for Grade 2 and Grade 3 fasteners. Higher grades can tolerate higher torque values.

If fasteners are laid out in a pattern - such as cylinder head bolts, oil pan bolts, differential cover bolts, etc. - loosen and tighten them in sequence to avoid warping the component. Where it matters, we'll show you this sequence. If a specific pattern isn't that important, the following rule-of-thumb guide will prevent warping.

First, install the bolts or nuts finger-tight. Then tighten them one full turn each, in a criss-cross or diagonal pattern. Then return to the first one and, following the same pattern, tighten them all one-half turn. Finally, tighten each of them one-quarter turn at a time until each fastener has been tightened to the proper torque. To loosen and remove the fasteners, reverse this procedure.

Metric thread sizes	Ft-lbs	Nm
M-6	6 to 9	9 to 12
M-8	14 to 21	19 to 28
M-10	28 to 40	38 to 54
M-12	50 to 71	68 to 96
M-14	80 to 140	109 to 154

Pipe thread sizes		
1/8	5 to 8	7 to 10
1/4	12 to 18	17 to 24
3/8	22 to 33	30 to 44
1/2	25 to 35	34 to 47

U.S. thread sizes		
1/4 – 20	6 to 9	9 to 12
5/16 – 18	12 to 18	17 to 24
5/16 – 24	14 to 20	19 to 27
3/8 – 16	22 to 32	30 to 43
3/8 – 24	27 to 38	37 to 51
7/16 – 14	40 to 55	55 to 74
7/16 – 20	40 to 60	55 to 81
1/2 – 13	55 to 80	75 to 108

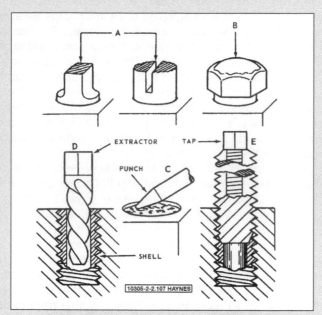

2.107 There are several ways to remove a broken fastener

A   *File it flat or slot it*
B   *Weld on a nut*
C   *Use a punch to unscrew it*
D   *Use a screw extractor (like an E-Z-Out)*
E   *Use a tap to remove the shell*

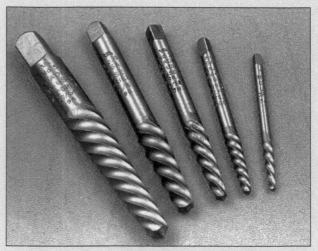

2.108 Typical assortment of E-Z-Out extractors

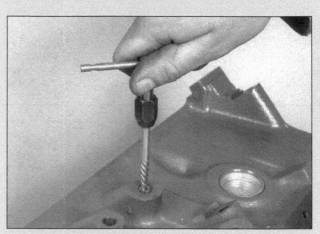

2.109 When screwing in the E-Z-Out, make sure it's centered properly

How to remove broken fasteners

Sooner or later, you're going to break off a bolt inside its threaded hole. There are several ways to remove it. Before you buy an expensive extractor set, try some of the following cheaper methods first.

First, regardless of which of the following methods you use, be sure to use penetrating oil. Penetrating oil is a special light oil with excellent penetrating power for freeing dirty and rusty fasteners. But it also works well on tightly torqued broken fasteners.

If enough of the fastener protrudes from its hole and if it isn't torqued down too tightly, you can often remove it with Vise-grips or a small pipe wrench. If that doesn't work, or if the fastener doesn't provide sufficient purchase for pliers or a wrench, try filing it down to take a wrench, or cut a slot in it to accept a screwdriver **(see illustration).** If you still can't get it off - and you know how to weld - try welding a flat piece of steel, or a nut, to the top of the broken fastener. If the fastener is broken off flush with - or below - the top of its hole, try tapping it out with a small, sharp punch. If that doesn't work, try drilling out the broken fastener with a bit only slightly smaller than the inside diameter of the hole. For example, if the hole is 1/2-inch in diameter, use a 15/32-inch drill bit. This leaves a shell which you can pick out with a sharp chisel.

If THAT doesn't work, you'll have to resort to some form of screw extractor, such as an E-Z-Out **(see illustration).**

Screw extractors are sold in sets which can remove anything from 1/4-inch to 1-inch bolts or studs. Most extractors are fluted and tapered high-grade steel. To use a screw extractor, drill a hole slightly smaller than the O.D. of the extractor you're going to use (Extractor sets include the manufacturer's recommendations for what size drill bit to use with each extractor size). Then screw in the extractor **(see illustration)** and back it - and the broken fastener - out. Extractors are reverse-threaded, so they won't unscrew when you back them out.

A word to the wise: Even though an E-Z-Out will usually save your bacon, it can cause even more grief if you're careless or sloppy. Drilling the hole for the extractor off-center, or using too small, or too big, a bit for the size of the fastener you're removing will only make things worse. So be careful!

How to repair broken threads

Sometimes, the internal threads of a nut or bolt hole can become stripped, usually from overtightening. Stripping threads is an all-too-common occurrence, especially when working with aluminum parts, because aluminum is so soft that it easily strips out. Overtightened spark plugs are another common cause of stripped threads.

Usually, external or internal threads are only partially stripped. After they've been cleaned up with a tap or die, they'll still work. Sometimes, however, threads are badly damaged. When this happens, you've got three choices:

1) *Drill and tap the hole to the next suitable oversize and install a larger diameter bolt, screw or stud.*

2) *Drill and tap the hole to accept a threaded plug, then drill and tap the plug to the original screw size. You can also buy a plug already threaded to the original size. Then you simply drill a hole to the specified size, then run the threaded plug into the hole with a bolt and jam nut. Once the plug is fully seated, remove the jam nut and bolt.*

3) *The third method uses a patented thread repair kit like Heli-Coil or Slimsert. These easy-to-use kits are designed to repair damaged*

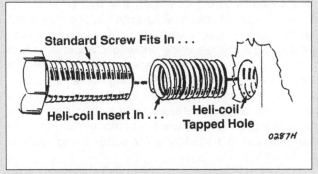

2.110 To install a Heli-Coil, drill out the hole, tap it with the special included tap and screw in the Heli-Coil

threads in spark plug holes, straight-through holes and blind holes. Both are available as kits which can handle a variety of sizes and thread patterns. Drill the hole, then tap it with the special included tap. Install the Heli-Coil **(see illustration)** *and the hole is back to its original diameter and thread pitch.*

Regardless of which method you use, be sure to proceed calmly and carefully. A little impatience or carelessness during one of these relatively simple procedures can ruin your whole day's work and cost you a bundle if you wreck an expensive head or block.

Component disassembly

Disassemble components carefully to help ensure that the parts go back together properly. Note the sequence in which parts are removed. Make note of special characteristics or marks on parts that can be installed more than one way, such as a grooved thrust washer on a shaft. It's a good idea to lay the disassembled parts out on a clean surface in the order in which you removed them. It may also be helpful to make sketches or take instant photos of components before removal.

When you remove fasteners from a component, keep track of their locations. Thread a bolt back into a part, or put the washers and nut back on a stud, to prevent mix-ups later. If that isn't practical, put fasteners in a fishing tackle box or a series of small boxes. A cupcake or muffin tin, or an egg crate, is ideal for this purpose - each cavity can hold the bolts and nuts from a particular area (i.e. oil pan bolts, valve cover bolts, engine mount bolts, etc.). A pan of this type is helpful when working on assemblies with very small parts, such as the carburetor or valve train. Mark each cavity with paint or tape to identify the contents.

When you unplug the connector(s) between two wire harnesses, or even two wires, it's a good idea to identify the two halves with numbered pieces of masking tape - or a pair of matching pieces of colored electrical tape - so they can be easily reconnected.

Gasket sealing surfaces

Gaskets seal the mating surfaces between two parts to prevent lubricants, fluids, vacuum or pressure from leaking out between them. Gaskets are often coated with a liquid or paste-type gasket sealing compound before assembly. Age, heat and pressure can cause the two parts to stick together so tightly that they're difficult to separate. Often, you can loosen the assembly by striking it with a soft-face hammer near the mating surfaces. You can use a regular hammer if you place a block of wood between the hammer and the part, but don't hammer on cast or delicate parts that can be easily damaged. When a part refuses to come off, look for a fastener that you forgot to remove.

Don't use a screwdriver or prybar to pry apart an assembly. It can easily damage the gasket sealing surfaces of the parts, which must be smooth to seal properly. If prying is absolutely necessary, use an old broom handle or a section of hard-wood dowel.

Once the parts are separated, carefully scrape off the

old gasket and clean the gasket surface. You can also remove some gaskets with a wire brush. If some gasket material refused to come off, soak it with rust penetrant or treat it with a special chemical to soften it, then scrape it off. You can fashion a scraper from a piece of copper tubing by flattening and sharpening one end. Copper is usually softer than the surface being scraped, which reduces the likelihood of gouging the part. The mating surfaces must be clean and smooth when you're done. If the gasket surface is gouged, use a gasket sealer thick enough to fill the scratches when you reassemble the components. For most applications, use a non-drying (or semi-drying) gasket sealer.

Hose removal tips

Warning: *If the vehicle is equipped with air conditioning, do not disconnect any of the A/C hoses without first having the system depressurized by a dealer service department or a service station* (see the Haynes Automotive Heating and Air Conditioning Manual).

The same precautions that apply to gasket removal also apply to hoses. Avoid scratching or gouging the surface against which the hose mates, or the connection may leak. Take, for example, radiator hoses. Because of various chemical reactions, the rubber in radiator hoses can bond itself to the metal spigot over which the hose fits. To remove a hose, first loosen the hose clamps that secure it to the spigot. Then, with slip-joint (or Channel lock) pliers, grab the hose at the clamp and rotate it around the spigot. Work it back and forth until it is completely free, then pull it off. Silicone or other lubricants will ease removal if they can be applied between the hose and the outside of the spigot. Apply the same lubricant to the inside of the hose and the outside of the spigot to simplify installation. Snap-On and Mac Tools sell hose removal tools - they look like bent ice picks - which can be inserted between the spigot and the radiator hose to break the seal between rubber and metal.

As a last resort - or if you're planning to replace the hose anyway - slit the rubber with a knife and peel the hose from the spigot. Make sure you don't damage the metal connection.

If a hose clamp is broken or damaged, don't reuse it. Wire-type clamps usually weaken with age, so it's a good idea to replace them with screw-type clamps whenever a hose is removed.

Automotive chemicals and lubricants

A wide variety of automotive chemicals and lubricants - ranging from cleaning solvents and degreasers to lubricants and protective sprays for rubber, plastic and vinyl - is available.

Cleaners

Brake system cleaner

Brake system cleaner removes grease and brake fluid from brake parts like disc brake rotors, where a spotless surface is essential. It leaves no residue and often eliminates brake squeal caused by contaminants. Because it leaves no residue, brake cleaner is often used for cleaning engine parts as well.

Carburetor and choke cleaner

Carburetor and choke cleaner is a strong solvent for gum, varnish and carbon. Most carburetor cleaners leave a dry-type lubricant film which will not harden or gum up. So don't use carb cleaner on electrical components.

Degreasers

Degreasers are heavy-duty solvents used to remove grease from the outside of the engine and from chassis components. They're usually sprayed or brushed on. Depending on the type, they're rinsed off either with water or solvent.

Demoisturants

Demoisturants remove water and moisture from electrical components such as alternators, voltage regulators, electrical connectors and fuse blocks. They are non-conductive, non-corrosive and non-flammable.

Electrical cleaner

Electrical cleaner removes oxidation, corrosion and carbon deposits from electrical contacts, restoring full current flow. It can also be used to clean spark plugs, carburetor jets, voltage regulators and other parts where an oil-free surface is necessary.

Lubricants

Assembly lube

Assembly lube is a special extreme pressure lubricant, usually containing moly, used to lubricate high-load parts (such as main and rod bearings and cam lobes) for initial start-up of a new engine. The assembly lube lubricates the parts without being squeezed out or washed away until the engine oiling system begins to function.

Graphite lubricants

Graphite lubricants are used where oils cannot be used due to contamination problems, such as in locks. The dry graphite will lubricate metal parts while remaining uncontaminated by dirt, water, oil or acids. It is electrically conductive and will not foul electrical contacts in locks such as the ignition switch.

Heat-sink grease

Heat-sink grease is a special electrically non-conductive grease that is used for mounting electronic ignition modules where it is essential that heat is transferred away from the module.

Moly penetrants

Moly penetrants loosen and lubricate frozen, rusted and corroded fasteners and prevent future rusting or freezing.

Motor oil

Motor oil is the lubricant formulated for use in engines. It normally contains a wide variety of additives to prevent corrosion and reduce foaming and wear. Motor oil comes in various weights (viscosity ratings) from 5 to 80. The recommended weight of the oil depends on the season, temperature and the demands on the engine. Light oil is used in cold climates and under light load conditions. Heavy oil is used in hot climates and where high loads are encountered. Multi-viscosity oils are designed to have characteristics of both light and heavy oils and are available in a number of weights from 5W-20 to 20W-50. Some home mechanics use motor oil as an assembly lube, but we don't recommend it, because motor oil has a relatively thin viscosity, which means it will slide off the parts long before the engine is fired up.

Silicone lubricants

Silicone lubricants are used to protect rubber, plastic, vinyl and nylon parts.

Wheel bearing grease

Wheel bearing grease is a heavy grease that can withstand high loads and friction, such as wheel bearings, balljoints, tie-rod ends and universal joints. It's also sticky enough to hold parts like the keepers for the valve spring retainers in place on the valve stem when you're installing the springs.

White grease

White grease is a heavy grease for metal-to-metal applications where water is present. It stays soft under both low and high temperatures (usually from -100 to +190-degrees F), and won't wash off or dilute when exposed to water. Another good "glue" for holding parts in place during assembly.

Sealants

Anaerobic sealant

Anaerobic sealant is much like RTV in that it can be used either to seal gaskets or to form gaskets by itself. It remains flexible, is solvent resistant and fills surface imperfections. The difference between an anaerobic sealant and an RTV-type sealant is in the curing. RTV cures when exposed to air, while an anaerobic sealant cures only in the absence of air. This means that an anaerobic sealant cures only after the assembly of parts, sealing them together.

RTV sealant

RTV sealant is one of the most widely used gasket compounds. Made from silicone, RTV is air curing, it seals, bonds, waterproofs, fills surface irregularities, remains flexible, doesn't shrink, is relatively easy to remove, and is used as a supplementary sealer with almost all low and medium temperature gaskets.

Thread and pipe sealant

Thread and pipe sealant is used for sealing hydraulic and pneumatic fittings and vacuum lines. It is usually made from a Teflon compound, and comes in a spray, a paint-on liquid and as a wrap-around tape.

Chemicals

Anaerobic locking compounds

Anaerobic locking compounds are used to keep fasteners from vibrating or working loose and cure only after installation, in the absence of air. Medium strength locking compound is used for small nuts, bolts and screws that may be removed later. High-strength locking compound is for large nuts, bolts and studs which aren't removed on a regular basis.

Anti-seize compound

Anti-seize compound prevents seizing, galling, cold welding, rust and corrosion in fasteners. High-temperature anti-seize, usually made with copper and graphite lubricants, is used for exhaust system and exhaust manifold bolts.

Gas additives

Gas additives perform several functions, depending on their chemical makeup. They usually contain solvents that help dissolve gum and varnish that build up on carburetor, fuel injection and intake parts. They also serve to break down carbon deposits that form on the inside surfaces of the combustion chambers. Some additives contain upper cylinder lubricants for valves and piston rings, and others contain chemicals to remove condensation from the gas tank.

Oil additives

Oil additives range from viscosity index improvers to chemical treatments that claim to reduce internal engine friction. It should be noted that most oil manufacturers caution against using additives with their oils.

Safety first!

Essential DOs and DON'Ts

Regardless of how enthusiastic you may be about getting on with the job at hand, take the time to ensure that your safety is not jeopardized. A moment's lack of attention can result in an accident, as can failure to observe certain simple safety precautions. The possibility of an accident will always exist, and the following points should not be considered a comprehensive list of all dangers. Rather, they are intended to make you aware of the risks and to encourage a safety-conscious approach to all work you carry out on your vehicle.

DON'T rely on a jack when working under the vehicle. Always use approved jackstands to support the weight of the vehicle and place them under the recommended lift or support points.

DON'T attempt to loosen extremely tight fasteners (i.e. wheel lug nuts) while the vehicle is on a jack - it may fall.

DON'T start the engine without first making sure that the transmission is in Neutral (or Park where applicable) and the parking brake is set.

DON'T remove the radiator cap from a hot cooling system - let it cool or cover it with a cloth and release the pressure gradually.

DON'T attempt to drain the engine oil until you are sure it has cooled to the point that it will not burn you.

DON'T touch any part of the engine or exhaust system until it has cooled sufficiently to avoid burns.

DON'T siphon toxic liquids such as gasoline, antifreeze and brake fluid by mouth, or allow them to remain on your skin.

DON'T inhale brake lining or clutch disc dust - it is potentially hazardous (see Asbestos below)

DON'T allow spilled oil or grease to remain on the floor - wipe it up before someone slips on it.

DON'T use loose-fitting wrenches or other tools which may slip and cause injury.

DON'T push on wrenches when loosening or tightening nuts or bolts. Always try to pull the wrench toward you. If the situation calls for pushing the wrench away, push with an open hand to avoid scraped knuckles if the wrench should slip.

DON'T attempt to lift a heavy component alone - get someone to help you.

DON'T rush or take unsafe shortcuts to finish a job.

DON'T allow children or animals in or around the vehicle while you are working on it.

DO wear eye protection when using power tools such as a drill, sander, bench grinder, etc. and when working under a vehicle.

DO keep loose clothing and long hair well out of the way of moving parts.

DO make sure that any hoist used has a safe working load rating adequate for the job.

DO get someone to check on you periodically when working alone on a vehicle.

DO carry out work in a logical sequence and make sure that everything is correctly assembled and tightened.

DO keep chemicals and fluids tightly capped and out of the reach of children and pets.

DO remember that your vehicle's safety affects that of yourself and others. If in doubt on any point, get professional advice.

Asbestos

Certain friction, insulating, sealing, and other products - such as brake linings, brake bands, clutch linings, torque converters, gaskets, etc. - contain asbestos. Extreme care must be taken to avoid inhalation of dust from such products since it is hazardous to health. If in doubt, assume that they do contain asbestos.

Batteries

Never create a spark or allow a bare light bulb near a battery. They normally give off a certain amount of hydrogen gas, which is highly explosive.

Always disconnect the battery ground (-) cable at the battery before working on the fuel or electrical systems.

If possible, loosen the filler caps or cover when charging the battery from an external source (this does not apply to sealed or maintenance-free batteries). Do not charge at an excessive rate or the battery may burst.

Take care when adding water to a non maintenance-free battery and when carrying a battery. The electrolyte, even when diluted, is very corrosive and should not be allowed to contact clothing or skin.

Always wear eye protection when cleaning the battery to prevent the caustic deposits from entering your eyes.

Fire

We strongly recommend that a fire extinguisher suitable for use on fuel and electrical fires be kept handy in the garage or workshop at all times. Never try to extinguish a fuel or electrical fire with water. Post the phone number for the nearest fire department in a conspicuous location near the phone.

Fumes

Certain fumes are highly toxic and can quickly cause unconsciousness and even death if inhaled to any extent. Gasoline vapor falls into this category, as do the vapors from some cleaning solvents. Any draining or pouring of such volatile fluids should be done in a well ventilated area.

When using cleaning fluids and solvents, read the instructions on the container carefully. Never use materials from unmarked containers.

Never run the engine in an enclosed space, such as a garage. Exhaust fumes contain carbon monoxide, which is extremely poisonous. If you need to run the engine, always do so in the open air, or at least have the rear of the vehicle outside the work area.

Gasoline

Remember at all times that gasoline is highly

flammable. Never smoke or have any kind of open flame around when working on a vehicle. But the risk does not end there. A spark caused by an electrical short circuit, by two metal surfaces contacting each other, or even by static electricity built up in your body under certain conditions, can ignite gasoline vapors, which, in a confined space, are highly explosive. Do not, under any circumstances, use gasoline for cleaning parts. Use an approved safety solvent. Also, DO NOT STORE GASOLINE IN A GLASS CONTAINER - use an approved metal or plastic container only!

Always disconnect the battery ground (-) cable at the battery before working on any part of the fuel system or electrical system. Never risk spilling a fuel on a hot engine or exhaust component.

Household current

When using an electric power tool, inspection light, etc., which operates on household current, always make sure that the tool is correctly connected to its plug and that, where necessary, it is properly grounded. Do not use such items in damp conditions and, again, do not create a spark or apply excessive heat in the vicinity of fuel or fuel vapor.

Secondary ignition system voltage

A severe electric shock can result from touching certain parts of the ignition system (such as the spark plug wires) when the engine is running or being cranked, particularly if components are damp or the insulation is defective. In the case of an electronic ignition system, the secondary system voltage is much higher and could prove fatal.

Keep it clean

Get in the habit of taking a regular look around the shop to check for potential dangers. Keep the work area clean and neat. Sweep up all debris and dispose of it as soon as possible. Don't leave tools lying around on the floor.

Be very careful with oily rags. Spontaneous combustion can occur if they're left in a pile, so dispose of them properly in a covered metal container.

Check all equipment and tools for security and safety hazards (like frayed cords). Make necessary repairs as soon as a problem is noticed - don't wait for a shelf unit to collapse before fixing it.

Accidents and emergencies

Shop accidents range from minor cuts and skinned knuckles to serious injuries requiring immediate medical attention. The former are inevitable, while the latter are, hopefully, avoidable or at least uncommon. Think about what you would do in the event of an accident. Get some first aid training and have an adequate first aid kit somewhere within easy reach.

Think about what you would do if you were badly hurt and incapacitated. Is there someone nearby who could be summoned quickly? If possible, never work alone just in case something goes wrong.

If you had to cope with someone else's accident, would you know what to do? Dealing with accidents is a large and complex subject, and it's easy to make matters worse if you have no idea how to respond. Rather than attempt to deal with this subject in a superficial manner, buy a good First Aid book and read it carefully. Better yet, take a course in First Aid at a local junior college.

Environmental safety

Be absolutely certain that all materials are properly stored, handled and disposed of. Never pour used or leftover oil, solvents or antifreeze down the drain or dump them on the ground. Also, don't allow volatile liquids to evaporate - keep them in sealed containers. Air conditioning refrigerant should never be expelled into the atmosphere. Have a properly equipped shop discharge and recharge the system for you.

3 Diagnosing engine problems

General information

This Chapter is devoted to engine checks and diagnosis. Correct diagnosis is an essential part of every repair; without it you can only cure the problem by accident.

Sometimes a simple tune-up item will cause symptoms similar to a worn out or defective engine. Be sure the engine is tuned to manufacturer's specifications before you begin with the following diagnosis procedures.

The tests and checks provided in this Chapter should help you decide if your engine needs a major overhaul. The information ranges from advice concerning oil consumption and loss of performance to detailed, step-by-step procedures covering spark plug reading, vacuum gauge and oil pressure checks, troubleshooting engine noises, measuring camshaft and timing chain wear to compression and leakdown tests.

If you are concerned about the condition of your engine because of decreased performance or fuel economy, perform a vacuum test, a power balance test and, if indicated, a compression test. If the engine passes these checks, inspect the spark plugs. Engines with high mileage and/or poor performance should also have the camshaft lobe lift and timing chain slack checked. If the engine is making unusual noises, perform an oil pressure test and noise diagnosis. Computer controlled engines may be checked with a scan tool to eliminate that system as the source of the problem.

Finally, after we have determined whether an overhaul or replacement engine is needed, we'll discuss the pros and cons of buying used or rebuilt engines or doing an overhaul yourself.

Reading symptoms

It's not always easy to determine when, or if, an engine should be completely overhauled, as a number of factors must be considered. Loss of power, rough running, knocking or metallic engine noises, excessive valve train noise and high fuel consumption rates may all point to the need for an overhaul, especially if they're all present at the same time.

High mileage is not necessarily an indication that an overhaul is needed, while low mileage doesn't preclude the need for an overhaul. Frequency of servicing is probably the most important consideration. An engine that's had regular and frequent oil and filter changes, as well as other required maintenance, will most likely give many thousands of miles of reliable service. Conversely, a neglected engine may require an overhaul very early in its life.

One of the more common reasons people rebuild their engines is because of oil consumption. Before you decide that the engine needs an overhaul based on oil consumption, make sure that oil leaks aren't responsible. If you park the vehicle in the same place everyday on pavement, look for oil stains or puddles of oil. To check more accurately, place a large piece of cardboard under the engine overnight. Compare the color and feel of the oil on the dipstick to the fluids found under the vehicle to verify that it's oil (transmission fluid is slightly red).

If any drips are evident, put the vehicle on a hoist and carefully inspect the underside. Sometimes leaks will only occur when the engine is hot, under load or on an incline, so look for signs of leakage as well as active drips. If a significant oil leak is found, correct it before you take oil consumption measurements.

Measuring oil consumption

Excessive oil consumption is an indication that cylinders, pistons, rings, valve stems, seals and/or valve guides may be worn. A clogged crankcase ventilation system can also cause this problem.

Every engine uses oil at a different rate. However, if an engine uses a quart of oil in 700 miles or less, or has visible blue exhaust smoke, it definitely needs repair.

To measure oil consumption accurately, park the vehicle on a level surface and shut off the engine. Wait about 15 minutes to allow the oil to drain down into the sump. Wipe the dipstick and insert it into the dipstick tube until it hits the stop. Then withdraw it carefully and read the level before the oil has a chance to flow. Fill the sump exactly to the full mark with the correct grade and viscosity of oil and write down the mileage shown on the odometer. Then monitor the oil level, using this same checking procedure until one quart of oil has been consumed and note the mileage.

Diagnostic checks

Internal noises

Every moving part in the vehicle can make noise. Owners frequently blame the engine for a noise that is actually in the transmission or driveline.

Set the parking brake and place the transmission in neutral (Park on automatics). Start the engine with the hood open and determine if the noise is actually coming from the engine. Rev the engine slightly; does the noise increase directly with engine speed? If the noise sounds like it's coming from the engine and it varies with engine speed, it probably is an engine or driveplate/torque converter noise (automatic transmissions only).

What kind of noise is it? If it's a squealing sound, check drivebelt tension. Spray some belt dressing (available at auto parts stores) on the belt(s); if the noise goes away, adjust or replace the belt(s) as necessary. Sometimes it's necessary to remove the drivebelt briefly to eliminate the engine accessories as a source of noise. With the belt removed and the engine off, turn each accessory by hand and listen for sounds. Then run the engine briefly and listen for the noise.

Determine if the sound occurs at crankshaft speed or half of crankshaft speed. If you are unsure, connect a timing light to any spark plug wire and listen while the light flashes. If the sound occurs every time the light flashes, it's occurring at half crankshaft speed. If the noise is audible twice for every flash, it's crankshaft speed. **Caution:** *Stay clear of moving engine components while the engine is running.*

Knocking or ticking noises are the most common types of internal engine noises. Noises that occur at crankshaft speed are usually caused by crankshaft, connecting rod and bearing problems, so begin your search on the lower part of the engine. Noises that occur at half of crankshaft

3.1 The best way to pinpoint the location of engine noises is with a mechanic's stethoscope

speed usually involve the camshaft, lifters, rocker arms, valves, springs and mechanical fuel pump pushrod. Listen for these sounds near the top of the engine.

To pinpoint the source of the noise, a mechanic's stethoscope is best **(see illustration)**. If you don't have one, improvise with a four-foot length of hose held to your ear. You can also hold the handle of a large screwdriver to your ear and touch the tip to the suspected areas.

Move the listening device around until the sound is loudest. Briefly short out the spark plug of the nearest cylinder and note how that affects the sound. Think about what components are in the area of the noise and which part could produce this sound.

If the noise is in the top of the engine, remove the valve cover from the affected side and briefly start the engine, allowing it to idle slowly so it doesn't fling too much oil. See if all the valves appear to be opening the same amount and if the pushrods are rotating slowly, as they should. Press down on each rocker arm with your thumb right above the valve. If the sound changes or goes away, you've found which cylinder has the problem. Check the valve clearance (see Chapter 7), then remove the rocker arm and pushrod. Inspect them carefully for wear and cracks. If no other problems are found, the lifter is probably defective. Check for lobe lift (discussed later in this Chapter) and replace the lifters (and camshaft) as necessary.

If the noise is in the lower portion of the engine, check for piston slap, wrist pin, connecting rod, main bearing and piston ring noises.

Piston slap is loudest on a cold engine and quiets down as the engine warms up. Listen for a dull, hollow sound in the cylinder wall just below the head that goes away or lessens if the spark plug is shorted out. Another way you can check for it is by slowly retarding the ignition timing while you listen. If it goes away or lessens, you've found the culprit.

If you suspect that a piston has a hole in it, remove the dipstick and listen in the opening. Hold a piece of rubber hose between your ear and the tube. You should be able to hear and feel combustion gasses escaping if the piston has a hole or is cracked.

Wrist pins make a double click that's quite easily heard at idle or low speeds. You can often tell which one is noisy by shorting out the appropriate spark plug.

Main bearing knocks have a low-pitched knock deep within the engine that will sound loudest when the engine is first started. It will also be quite noticeable under heavy load. You can often tell which one is noisy by shorting out the adjacent spark plugs.

Connecting rod bearings knock loudest when the accelerator is pressed briefly and then quickly released. The noise is usually caused by excessive bearing wear or insufficient oil pressure.

Piston rings that are loose in their grooves or broken make a chattering noise that is loudest during acceleration. This should be confirmed by a cylinder leak-down test.

Reading spark plugs

General information

The spark plugs provide a sort of window into the combustion chamber and can give a wealth of information about engine operation to a savvy mechanic. Fuel mixture, heat range, oil consumption and detonation all leave their mark on the tips of the spark plugs.

Before you begin the check, drive the vehicle at highway speed, allowing it to warm up thoroughly without excessive idling. Shut the engine off and wait until it cools sufficiently so you won't get burned if you touch the exhaust manifolds.

Check the spark plug wires to see if they have the cylinder numbers on them. Label them if necessary so you can reinstall them on the correct spark plugs.

Remove the spark plugs and place them in order on top of the air cleaner. Note the brand and number on the plugs **(see illustration).** Compare this to the factory recom-

3.2 **Note the brand and number on the spark plugs**

mendations to determine if the correct type and heat range was used.

Heat range

Spark plug manufacturers make spark plugs in several heat ranges for different driving conditions. Engines that are run hard produce more heat in the combustion chamber than engines that are driven mostly at slow speeds. Plugs with a colder heat range are designed to transfer heat away from the tip more rapidly than hotter spark plugs. As a general rule, if most of your driving is low speed, around town short trips, use a hotter heat range to help burn off any deposits that may form on the plug tip. If the majority of your driving is done at higher speeds, use a colder heat range to prevent overheating of the tip.

The engine must have the correct heat range spark plugs before you can read the tips accurately. Plugs that are too hot will mask a rich fuel mixture reading; conversely, cold plugs will tend to foul on a normal mixture. On most European and Japanese spark plugs, the higher the number, the colder the heat range. American plugs are just the opposite.

There are several "old wives tales" about heat range that need to be dispelled. Hotter heat range plugs don't make the engine run hotter, they don't make a hotter spark and they don't increase combustion chamber temperature (unless the colder plug wasn't firing).

Perhaps the best way for a home mechanic to tell if the heat range is correct is to look at the center electrode tip. There should be a slightly bluish ring around the electrode between the tip and insulator cone. If the fuel mixture is too rich, it may be necessary to scrape some carbon off to see this.

If the plugs are too hot, the electrodes may be burned away, the insulators may be blistered or bubbled, or deposits on the insulator may be glazed or melted. If the plugs are too cold, the porcelain tip is probably going to be sooty with black carbon deposits.

The factory recommended heat ranges are the best starting point for stock or slightly modified engines. Usually a change of not more than one heat range is all that's needed to tailor the plugs to your driving conditions.

Reading plugs for diagnosis

Now that you have the right heat range, you can examine the plugs for hints about the engine's internal condition and state of tune. If any of the plugs are wet with oil, engine repairs are needed right away. If the plugs have significant gray or white deposits, it means that a moderate amount of oil is seeping into the cylinders and repairs will be needed soon, or you've been doing a lot of short trip driving.

The ideal color for plugs used in engines run on leaded gasoline is light brown on the insulator cone and beige on the ground (side) electrode. Engines run on unleaded gasoline tend to leave very little color on the plugs. Late-model emission-controlled engines run very lean. Normally, the plugs range from almost white to tan on the porcelain insulator cone and the ground electrode should be light brown to dark gray.

Excessively rich fuel mixtures cause the spark plugs tips to turn black and lean mixtures result in light tan or white tips. You can tell by the color if the fuel mixture is in the ballpark by reading the plugs, but to be sure you have the right fuel mixture. Run the engine on an exhaust gas analyzer connected ahead of the catalytic converter (if equipped).

If the engine has a misfire and one or more plugs are carbon fouled, look for an ignition problem or low compression in the affected cylinder(s). Sometimes the spark plugs will vary among each other in color because of improper mixture distribution. Look for a leaky intake manifold gasket if one or more adjoining cylinders are running very lean. If the plugs are burning unevenly, one side of the carburetor may be faulty, you may have a vacuum leak, or, on fuel-injected models, you may have bad injectors.

Detonation, preignition and plugs that are too long can result in physical damage to the tip. Check the accompanying color photos **(see inside of back cover)** to help identify these problems.

Vacuum gauge diagnostic checks

A vacuum gauge provides valuable information about what is going on in the engine at a low cost. You can check for worn rings or cylinder walls, leaking head or intake manifold gaskets, incorrect carburetor adjustments, restricted exhaust, stuck or burned valves, weak valve springs, improper ignition or valve timing and ignition problems.

Unfortunately, vacuum gauge readings are easy to misinterpret, so they should be used in conjunction with other tests to confirm the diagnosis.

Both the absolute readings and the rate of needle movement are important for accurate interpretation. Most gauges measure vacuum in inches of mercury (in-Hg). The following typical vacuum gauge readings assume the diagnosis is being performed at sea level. As elevation increases (or atmospheric pressure decreases), the reading will decrease. For every 1,000 foot increase in elevation above approximately 2,000 feet, the gauge readings will decrease about one inch of mercury.

3.3 Connect the vacuum gauge directly to intake manifold vacuum

Connect the vacuum gauge directly to intake manifold vacuum, not to ported (carburetor) vacuum **(see illustration)**. Be sure no hoses are left disconnected during the test or false readings will result.

Before you begin the test, allow the engine to warm up completely. Block the wheels and set the parking brake. With the transmission in neutral (or Park, on automatics), start the engine and allow it to run at normal idle speed. **Warning:** *Carefully inspect the fan blades for cracks or damage before starting the engine. Keep your hands and the vacuum tester clear of the fan and do not stand in front of the vehicle or in line with the fan when the engine is running.*

Read the vacuum gauge; a healthy engine should produce about 15 to 20 inches of vacuum with a fairly steady needle. **Note:** *Engines with high performance camshafts will have a lower and somewhat erratic reading at idle. Emission controlled vehicles from the mid-1970s tend to have slightly lower vacuum due to modified valve timing and retarded ignition timing.* Refer to the following vacuum gauge readings and what they indicate about the engine:

0279 H

3.4 Low, steady reading

Low steady reading

This usually indicates a leaking gasket between the intake manifold and carburetor or throttle body, a leaky vacuum hose, late ignition timing or incorrect camshaft timing **(see illustration)**. Check ignition timing with a timing light and eliminate all other possible causes, utilizing the tests provided in this Chapter before you remove the timing chain cover to check the timing marks.

Low, fluctuating reading

If the needle fluctuates about three to eight inches below normal **(see illustration)**, suspect an intake manifold gasket leak at an intake port or a faulty injector (on port-injected models only).

3.5 Low, fluctuating needle

Regular drops

If the needle drops about two to four inches at a steady rate **(see illustration)**, the valves are probably leaking. Perform a compression or leakdown test to confirm this.

3.6 Regular drops

Irregular drops

An irregular down-flick of the needle **(see illustration)** can be caused by a sticking valve or an ignition misfire. Perform a compression or leakdown test and read the spark plugs.

3.7 Irregular drops

Rapid vibration

A rapid four in. Hg vibration at idle **(see illustration)** combined with exhaust smoke indicates worn valve guides. Perform a leakdown test to confirm this. If the rapid vibration occurs with an increase in engine speed, check for a leaking intake manifold gasket or head gasket, weak valve springs, burned valves or ignition misfire.

3.8 Rapid vibration

Slight fluctuation

A slight fluctuation, say one inch up and down, may mean ignition problems. Check all the usual tune-up items and, if necessary, run the engine on an ignition analyzer.

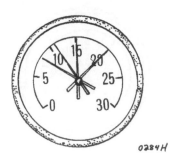

3.9 Large fluctuation

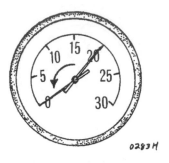

3.10 Slow return after revving

Large fluctuation

If this occurs (see illustration), perform a compression or leakdown test to look for a weak or dead cylinder or a blown head gasket.

Slow hunting

If the needle moves slowly through a wide range, check for a clogged PCV system, incorrect idle fuel mixture, carburetor/throttle body or intake manifold gasket leaks.

Slow return after revving

Quickly snap the throttle open until the engine reaches about 2,500 rpm and let it shut. Normally the reading should drop to near zero, rise above normal idle reading (about 5 in-Hg over) and then return to the previous idle reading (see illustration). If the vacuum returns slowly and doesn't peak when the throttle is snapped shut, the rings may be worn. If there is a long delay, look for a restricted exhaust system (often the muffler or catalytic converter). An easy way to check this is to temporarily disconnect the exhaust ahead of the suspected part and redo the test.

3.11a A gauge with a threaded fitting for the spark plug hole is preferred over the type that requires hand pressure to maintain the seal during the compression check

3.11b Connect a remote starter switch to the positive terminal of the battery and the small terminal on the starter solenoid, as shown here

Compression check

1   A compression check will tell you what mechanical condition the upper end (pistons, rings, valves, head gaskets) of your engine is in. Specifically, it can tell you if the compression is down due to leakage caused by worn piston rings, defective valves and seats or a blown head gasket.

Note: *The engine must be at normal operating temperature and the battery must be fully charged for this check. Also, if the engine is equipped with a carburetor, the choke valve must be all the way open to get an accurate compression reading (if the engine's warm, the choke should be open).*

2   Begin by cleaning the area around the spark plugs before you remove them (compressed air should be used, if available, otherwise a small brush or even a bicycle tire pump will work). The idea is to prevent dirt from getting into the cylinders as the compression check is being done.

3   Remove all of the spark plugs from the engine. Be sure

COMPRESSION TEST PERCENTAGE CHART

Maximum kPa (PSI)	Minimum kPa (PSI)	Maximum kPa (PSI)	Minimum kPa (PSI)	Maximum kPa (PSI)	Minimum kPa (PSI)	Maximum kPa (PSI)	Minimum kPa (PSI)
923.23 (134)	696.40 (101)	1130.78 (164)	848.09 (123)	1337.63 (194)	999.78 (145)	1544.48 (224)	1158.36 (168)
937.72 (136)	703.29 (102)	1144.57 (166)	858.98 (124)	1351.42 (196)	1013.57 (147)	1558.27 (226)	1165.26 (169)
951.51 (138)	717.08 (104)	1158.36 (168)	868.77 (126)	1365.21 (198)	1020.46 (148)	1572.06 (228)	1179.65 (171)
965.30 (140)	723.98 (105)	1172.15 (170)	875.67 (127)	1379.00 (200)	1034.25 (150)	1585.85 (230)	1185.94 (172)
979.09 (142)	737.77 (107)	1185.94 (172)	889.46 (129)	1392.79 (202)	1041.15 (151)	1599.64 (232)	1199.23 (174)
992.88 (144)	744.66 (108)	1199.73 (174)	903.25 (131)	1406.58 (204)	1054.94 (153)	1613.43 (234)	1206.63 (175)
1006.67 (146)	758.45 (110)	1206.63 (176)	910.14 (132)	1420.37 (206)	1061.83 (154)	1627.22 (236)	1220.42 (177)
1020.46 (148)	765.35 (111)	1227.31 (178)	917.04 (133)	1434.16 (208)	1075.62 (156)	1641.01 (238)	1227.31 (178)
1034.25 (150)	779.14 (113)	1241.10 (180)	930.83 (135)	1447.95 (210)	1082.52 (157)	1654.80 (240)	1241.10 (180)
1048.04 (152)	786.03 (114)	1254.89 (182)	937.72 (136)	1461.74 (212)	1089.41 (158)	1668.59 (242)	1248.00 (181)
1061.83 (154)	792.93 (115)	1268.68 (184)	951.51 (138)	1475.53 (214)	1103.20 (160)	1682.38 (244)	1261.79 (183)
1075.62 (156)	806.72 (117)	1282.47 (186)	965.30 (140)	1489.32 (216)	1116.99 (162)	1696.17 (246)	1268.68 (184)
1089.41 (158)	813.61 (118)	1296.26 (188)	972.20 (141)	1503.11 (218)	1123.89 (163)	1709.96 (248)	1282.47 (186)
1103.20 (160)	827.40 (120)	1310.05 (190)	979.09 (142)	1560.90 (220)	1137.68 (165)	1723.75 (250)	1289.37 (187)
1116.99 (162)	834.30 (121)	1323.84 (192)	992.88 (144)	1530.69 (222)	1144.57 (166)		

3.12 Locate your maximum compression reading on the chart and look to the right to find the minimum acceptable compression, then compare it to your lowest reading

to keep them in order so you can read the tips.

4   Block the throttle wide open.

5   Detach the coil wire from the center of the distributor cap and ground it on the engine block. Use a jumper wire with alligator clips on each end to ensure a good ground. On EFI-equipped vehicles, the fuel pump circuit should also be disabled by removing the fuse.

6   Install the compression gauge in the number one spark plug hole **(see illustration)**.

7   Crank the engine over at least five compression strokes with a remote starter switch and watch the gauge **(see illustration)**. The compression should build up quickly in a healthy engine. Low compression on the first stroke, followed by gradually increasing pressure on successive strokes, indicates worn piston rings. A low compression reading on the first stroke, which doesn't build up during successive strokes, indicates leaking valves or a blown head gasket (a cracked head could also be the cause). Deposits on the undersides of the valve heads can also cause low compression. Record the highest gauge reading obtained.

8   Repeat the procedure for the remaining cylinders. If readings very more than 20-percent among cylinders, internal engine components are excessively worn.

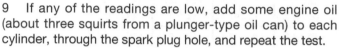

3.13a Remove the oil pressure sending unit (arrow) - it's either adjacent to the oil filter or on top of the block, behind the intake manifold

3.13b Screw a mechanical oil pressure gauge in place of the sending unit

9   If any of the readings are low, add some engine oil (about three squirts from a plunger-type oil can) to each cylinder, through the spark plug hole, and repeat the test.

10   If the compression increases after the oil is added, the piston rings are definitely worn. If the compression doesn't increase significantly, the leakage is occurring at the valves or head gasket. Leakage past the valves may be caused by burned valve seats and/or faces or warped, cracked or bent valves.

11   If two adjacent cylinders have equally low compression, there's a strong possibility that the head gasket between them is blown. The appearance of coolant in the combustion chambers or the crankcase would verify this condition.

12   If one cylinder is about 20 percent lower than the others, and the engine has a slightly rough idle and/or backfires, a worn lobe on the camshaft could be the cause.

13   If the compression is unusually high, the combustion chambers are probably coated with carbon deposits. If that's the case, the cylinder heads should be removed and decarbonized.

14   If compression is way down or varies greatly between cylinders, it would be a good idea to perform a leak-down test, as described later in this Chapter. This test will pinpoint exactly where the leakage is occurring and how severe it is.

Oil pressure check

Engine oil pressure provides a fairly good indication of bearing condition in an engine. As bearing surfaces wear, oil clearances increase. This increased clearance allows the oil to flow through the bearings more readily, which results in lower oil pressure. Oil pumps also wear, which causes an additional loss of pressure.

Pull out the dipstick and check the oil level. If the oil is dirty, contaminated with gasoline from short trips, or too thin (low viscosity) for the season, change it.

Check the oil pressure with a gauge installed in place

of the oil pressure sending unit **(see illustrations)**. Allow the engine to reach normal operating temperature before performing the test. If the pressure is extremely low, the bearings and/or oil pump are probably worn out. As a general rule, there should be about ten psi of pressure for each 1,000 rpm of engine speed.

Camshaft lobe lift check

Worn camshaft lobes are fairly common on high-mileage engines. If the engine is low on power, runs rough and/or backfires constantly through the intake or exhaust, suspect crossed ignition wires, a cracked distributor cap or a worn camshaft. These problems won't show up on a compression test (unless a valve doesn't open at all), but will show up on a power balance test.

To check for worn camshaft lobes, remove the valve covers. Then remove the rocker arms or rotate them aside from the pushrods.

Mount a dial indicator so it bears on the end of the pushrod **(see illustration)**.

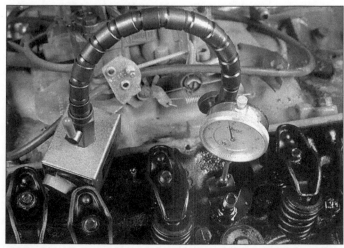

3.14 Mount a dial indicator so the pushrod will act directly on the plunger

3.15 Measure the distance to the chain when it's tight . . .

3.16 . . . then measure the difference due to slack

Using a socket and ratchet on the vibration damper bolt, slowly turn the crankshaft clockwise through two full rotations (720-degrees) while observing the dial indicator. Note the high and low readings and subtract the low from the high to obtain the lift. Record the measurements for each cylinder and note whether it was for exhaust or intake.

Compare the measurements of all the intake lobes; they should be within 0.005 inch of each other. Repeat this check for the exhaust lobes as well.

When the surface hardening wears off of a camshaft lobe, the metal below it scuffs away rapidly. Usually, worn lobes will be several tenths of an inch below the good ones.

If one or more lobes are worn, replace the camshaft and lifters as an assembly. Never use old lifters on a new camshaft.

Timing chain slack check

1   Remove the timing chain cover (see Chapter 4).
2   Temporarily reinstall the vibration damper bolt. Using this bolt, rotate the crankshaft in a counterclockwise direction to take up the slack in the right (passenger's) side of the chain.
3   Establish a reference point on the block and measure from that point to the chain **(see illustration)**.
4   Rotate the crankshaft in a clockwise direction to take up the slack in the left (driver's) side of the chain.
5   Force the left side of the chain out with your fingers and measure the distance between the reference points and the chain **(see illustration)**. The difference between the two measurements is the slack.
6   If the slack exceeds 1/2-inch, install a new chain and sprockets.

Power balance test

A power balance test is used to determine which cylinder(s) are not doing their share of the work; a condition that results in a rough and uneven idle, skipping and loss of power. The spark to each cylinder is shorted out or disconnected momentarily and a tachometer is used to measure the resultant drop in idle speed. If a cylinder is not producing power, the idle speed won't drop when it's disconnected. **Caution:** *On vehicles with electronic ignition, short the spark plug wire to ground to avoid damaging the components.*

Most electronic engine analyzers have a power balance test function which enables the operator to quickly check the difference among cylinders. If you don't have access to an analyzer, you may perform the same test using a tachometer, such as the kind found in dwell/tachs.

Connect a test tachometer in accordance to the manufacturer's instructions. Allow the engine to warm up completely. With the parking brake set and the transmission in neutral (or Park on automatics), record the idle speed.

Disconnect the spark plug wires one at a time and record the rpm drop for each one, then compare the results **(see illustration)**. On models with electronic ignition, immediately short the wire to an engine ground to prevent damage to the system. Little change in idle speed when a cylinder is shorted out indicates a weak cylinder. **Warning:** *Use a tool made of a non-conducting material (such as plastic) to disconnect the wires - the high voltage in the ignition sys-*

3.17 Disconnect the spark plug wires one at a time and note the rpm drop

tem can cause severe shocks. **Caution:** *If the engine is equipped with a catalytic converter, do not leave a plug disconnected or shorted out for more than 15 seconds. Then allow about 30 seconds for the converter to cool before testing the next cylinder.* **Note:** *Since wire boots often become stuck on the spark plugs over time, it's a good idea to remove and replace each wire from each plug, one at a time, with the engine off. This will prevent struggling with them while the engine is running and you're using the plastic tool. If the engine has an EGR valve, disconnect and plug the hose during the test.*

On a healthy engine, the variation in rpm drop will not be more than 40 or 50 between the highest and lowest cylinders. If any cylinders differ more than this, there is a problem.

If any cylinders fail the power balance test, perform a compression test or a leak-down test to determine if compression is satisfactory. If compression is low, perform the necessary repair operations to correct it. If compression is normal, look for ignition problems such as faulty spark plugs, spark plug wires or distributor cap. A power balance test is also affected by faulty fuel system components. Look for vacuum leaks in the gasket between the intake manifold and cylinder head. On port fuel-injected models, also check for an inoperative fuel injector.

Cylinder leak-down test

Any time a compression test identifies one or more weak cylinders, you should perform a cylinder leak-down test to determine exactly where the problem lies. A cylinder leak-down test will pinpoint problems like a blown head gasket, leaking intake or exhaust valves, cracked cylinder walls and heads and faulty pistons and rings. Since leak-down testers are expensive, you'll probably want to have this procedure performed by a shop that has the equipment.

Leak-down testers **(see illustration)** pressurize the cylinder with compressed air supplied by an air compressor. Air is forced into the cylinder with the valves closed and the rate of leakage is measured as a percentage.

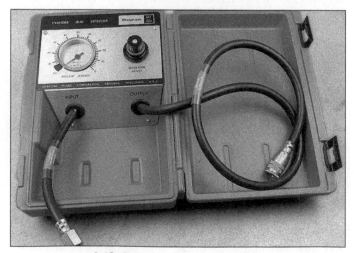

3.18 A typical leak-down tester

Prepare the engine for the test by removing all the spark plugs (label the wires with the cylinder numbers, if necessary). Test the cylinders one at a time, following the firing order. **Note:** *The firing order and cylinder numbers are normally stamped on the intake manifold.*

Disconnect the cable from the negative terminal of the battery. Beginning with the number one cylinder, bring the piston to Top Dead Center (TDC) on the compression stroke. Have an assistant turn the bolt in the center of the vibration damper clockwise slowly with a socket and breaker bar. Hold a finger over the spark plug opening to detect air pressure; this indicates the piston is on the compression stroke. **Note:** *Some testers come with a special "whistle" that performs this function. Always follow the instructions provided by the tester manufacturer.*

Stop turning the crankshaft when the "0" or TDC mark on the vibration damper aligns with the pointer on the timing chain cover.

Screw the air hose into the spark plug hole and connect the tester. Attach the line from the air compressor and zero the gauge by turning the pressure regulator knob. **Note:** *The crankshaft must be held in this position; when air pressure is applied, it will tend to move. On manual transmission models, engage high gear and set the parking brake. On automatic transmission models, have an assistant hold the crankshaft with a socket and long breaker bar on the bolt in the center of the vibration damper.* **Warning:** *Make sure the breaker bar is held securely, and the rest of the person's body is safely away from it - the bar could slip and cause injury!*

Check the leak-down rate. A rate of 10-percent is considered normal; if a cylinder leakage rate is 20-percent or more, it's rebuild time.

With the cylinder pressurized, air will leak past the worn/defective parts. By listening for the escaping air, you can tell where it's escaping from. If air is heard out the tail pipe, the exhaust valve is leaking and air coming from the intake manifold indicates a leaking intake valve. Remove the oil filler cap and listen for air sounds coming from the crankcase, which indicates blow-by past the pistons and rings.

Repeat this procedure for each cylinder following the firing order and record the results. Mark the vibration damper exactly at 90-degree (1/4-turn) intervals from the "O" or TDC mark. Then turn the crankshaft 1/4-turn (90-degrees) to get to TDC of the next cylinder in the firing order. If you're unsure about reaching TDC, insert a plastic soda straw into the spark plug hole (after air under pressure starts coming out) while the crankshaft is being turned. The point where the straw stops moving out and before moving back in is TDC.

Cooling system tests

A cracked block or cylinder head and/or a blown head gasket will cause a loss of power, overheating and a host of other problems. These problems are usually brought on by severe overheating or freezing due to insufficient antifreeze.

3.19 Pump up the cooling system tester until the pressure equals the rating on the radiator cap (usually about 15 psi), then look for leaks and watch the gauge on the tester for a pressure drop

3.20 Use the squeeze bulb to draw a sample into the tester

If you suspect internal engine leakage, check the oil on the dipstick for contamination of the oil by coolant. The oil level may increase and the oil will appear milky in color. Sometimes, oil will also get into the radiator; it will usually float on the top of the coolant. Occasionally steam and coolant will come out of the exhaust pipe, even when the engine is warmed up, because of a leaking head gasket. **Note:** *Don't confuse this with the condensation vapor normally present when an engine is warming up in cool weather).*

The cooling system and engine may be checked for leaks with a pressure tester **(see illustration)**. Follow the instructions provided by the tool manufacturer. Correct any external leaks in the hoses, water pump and radiator, etc. If no external leaks are found, look and listen for signs of leakage on the engine. **Note:** *A leaking heater core will cause a hidden pressure loss. Clamp off the hoses going to it to at the firewall to eliminate this source of leakage.*

Another device that is useful for determining if there is a crack in the engine or a blown head gasket is a combustion leak block tester. Combustion leak block testers **(see illustration)** use a blue colored fluid to test for combustion gases in the cooling system, which indicates a compression leak from a cylinder into the coolant. Be sure to follow the instructions included with the tester. A sample of gasses present in the top of the radiator is drawn into the tester. If any combustion gases are present in the sample taken, the test fluid will change color to yellow. Block testers and extra test fluid are readily available from most auto parts stores.

If the engine is overheating but testing indicates no cracks, blown gaskets or other internal problems, carefully inspect the cooling system for problems and correct as necessary. Frequently, partially clogged radiators, stuck thermostats and defective water pumps cause overheating. If any of the above tests indicate an internal engine problem, try to determine which cylinder(s) are affected by removing the spark plugs and checking the tips. If coolant

is getting into a combustion chamber, the plug will either be completely clean or will have traces of coolant on it. Sometimes coolant will even leak from the spark plug hole.

When an internal coolant leak is found, the cylinder heads should be removed for a thorough inspection. If a gasket has blown, have an automotive machine shop check for warpage on both heads and resurface as necessary. If no warpage is found, have both cylinder heads checked for cracks. If tests indicate internal leakage, but the heads check out OK, have the block checked.

Correct the cause of the failure, such as a clogged radiator, before the vehicle is put back in service. Otherwise, the problem will likely reoccur.

Scan tools

Computerized engines develop unique problems of their own that conventional testers can't always find. Aftermarket tool companies have developed a number of special testers, known generically as scan tools, to test computer controlled engines. These tools cause the computers to enter the diagnostic mode and measure various components and sub-systems electronically.

Most of these tools use prompts for data entry and provide encoded outputs that indicate the condition of each part in the system. If you have a computer controlled model and the "Check Engine" light is on, a scan tool of some kind will probably be required to properly diagnose the vehicle. See your local auto parts dealer for features and costs of the scan tools available to home mechanics.

Is the engine worth rebuilding?

The do-it-yourselfer is faced with a number of options when the time comes for an engine overhaul. The decision to replace the engine block, piston/connecting rod assemblies and crankshaft depends on a number of factors, with the number one consideration being the condition of the block. Other considerations are cost, access to machine shop facilities, parts availability, time required to complete the project and the extent of prior mechanical experience on the part of the do-it-yourselfer. Also, consider the value of the vehicle. Frequently, the cost of the parts to rebuild the engine is higher than the vehicle's value. In such a case, you may want to consider a used engine or ultimately junking the vehicle. A rebuilt engine doesn't raise the value of a vehicle as much as some people think.

Before beginning the engine overhaul, familiarize yourself with the scope and requirements of the job. Overhauling an engine isn't difficult if you follow all of the instructions carefully, have the necessary tools and equipment and pay close attention to all specifications; however, it is time consuming. Plan on the vehicle being tied up for a minimum of two weeks, especially if parts must be taken to an automotive machine shop for repair or reconditioning. Check on availability of parts and make sure that any necessary special tools and equipment are obtained in advance. Most work can be done with typical hand tools, although a number of precision measuring tools are required for inspecting parts to determine if they must be replaced. Often an automotive machine shop will handle the inspection of parts and offer advice concerning reconditioning and replacement.

Note: *Always wait until the engine has been completely disassembled and all components, especially the engine block, have been inspected before deciding what service and repair operations must be performed by an automotive machine shop. Since the block's condition will be the major factor to consider when determining whether to overhaul the original engine or buy a used or rebuilt one, never purchase parts or have machine work done on other components until the block has been thoroughly inspected. As a general rule, time is the primary cost of an overhaul, so it doesn't pay to install worn or substandard parts.*

Some of the rebuilding alternatives include:

Used engines - A used engine is usually the least expensive way to get a vehicle with a defective engine back on the road again. However, there are several drawbacks. If you purchase a used engine without hearing it run, it may be in poor condition. If you buy one as-is, you may lose the cost of the engine if it turns out to be bad. Some salvage yards provide a short warranty, but won't reimburse you for the labor required to replace a faulty engine you purchase from them. Additionally, a used engine is already partly

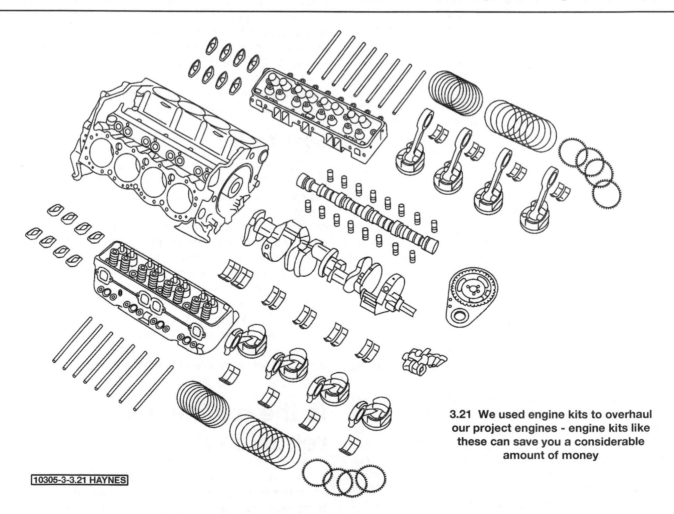

3.21 We used engine kits to overhaul our project engines - engine kits like these can save you a considerable amount of money

10305-3-3.21 HAYNES

worn out when you get it, and is more likely to fail at any time than a new or rebuilt unit.

Individual parts - If the inspection procedures reveal that the engine block and most engine components are in reusable condition, purchasing individual parts may be the most economical alternative. The cylinder heads, block, camshaft, crankshaft and piston/connecting rod assemblies should all be inspected carefully. Even if the block shows little wear, the cylinder bores should be surface honed and the piston rings replaced.

Many auto parts stores and mail order houses sell overhaul kits composed of bearings, seals, gaskets, rings, etc. **(see illustration)**. Buying such kits is usually less expensive than purchasing each item separately. Also, piston and crankshaft kits are available from the same sources. A lot of money can be saved by shopping. Compare prices before you buy.

Short block - A short block consists of an engine block with a crankshaft and piston/connecting rod assemblies already installed. All new bearings are incorporated and all clearances will be correct. Some short blocks come with camshafts and lifters; if not, the existing camshaft and lifters, along with the valve train components, cylinder heads and external parts can be bolted to the short block with little or no machine shop work necessary.

Long block - A long block consists of a short block plus an oil pump, oil pan, cylinder heads, camshaft and valve train components, timing sprockets and chain and timing cover. These engines are frequently called "remanufactured". All components are installed with new bearings, seals and gaskets incorporated throughout. The installation of manifolds and external parts is usually all that's necessary.

New replacement engines - Complete new engines are available through Chevrolet dealers. These cost more than other alternatives, but usually have the longest warranties and tend to last longer than rebuilt units.

Give careful thought to which alternative is best for you and discuss the situation with local automotive machine shops, auto parts dealers and experienced rebuilders before ordering or purchasing replacement parts.

Notes

4 Preparing for an overhaul

Preparing to remove the engine

Removing an engine is one of the most troublesome and potentially dangerous phases of rebuilding an engine. The degree of difficulty becomes compounded when it comes time to install the engine - you must remember how everything goes back together. A careful and orderly removal will minimize any problems and help the project to go smoothly.

There are several preliminary steps that should be taken before the engine can be removed.

Locating a place to work is extremely important. Adequate work space along with storage space for the vehicle will be needed. If a shop or garage is not available, at the very least a flat, level, clean work surface made of concrete is required. If the only surface available is asphalt, be sure to remember that most jackstands will tend to "sink" into the asphalt when they are supporting an extreme amount of weight. Use plywood or sheet metal under the jackstands to provide a rigid surface for the jackstands when they are supporting the vehicle.

Cleaning the engine compartment and the engine before beginning the removal procedure will help keep the tools clean and organized. The best method for cleaning the engine and engine compartment is to have it steam cleaned by a professional automobile detail shop. An alternate method most used by home mechanics is to have the engine degreased. Purchase a can of engine degreaser from the local auto parts store (two cans if the engine and compartment have an extraordinary amount of dirt and grease). There are several brands to choose from, but be sure the label specifically designates the product as an engine degreaser, not a brake or carburetor cleaner! Before the vehicle has been disassembled, drive it over to a self-service type car wash or to a place that is equipped with high-pressure cleaning equipment and proper drainage. Make sure the trip over to the car wash has allowed the engine to heat up to normal operating temperature to allow the degreaser to take "action" when it is applied. Use a few old rags or a plastic bag and wrap the distributor carefully to prevent any water from entering. This will help the engine to start after it has been cleaned. Spray the engine degreaser onto the engine and engine compartment walls. Concentrate on the areas where the grease has collected the most (e.g. the fuel pump area and the timing chain cover area, etc.). Allow the degreaser or solvent to penetrate the grease for five or ten minutes before continuing. Insert the coins into the machine, turn the selector to WASH and start cleaning the engine compartment first. Move the pressure nozzle in close to "blast" the areas that are layered with dirt and grease and back off the nozzle for cleaning sensitive parts such as relays, filters or any small electrical units. When cleaning any electrical parts with high-pressure hot water, use caution and personal judgement to make sure the parts are properly cleaned but not unintentionally damaged. Next, move the high-pressure nozzle over the engine. Start on the lower sections of the engine first, so that the water and soap will not cover over the grease and make it difficult to see the dirty areas. Work your way around the engine, carefully cleaning the corners of the engine as well as the more obvious surfaces of the timing chain cover and the valve covers. Finish the job by rinsing off the entire engine compartment and engine with the machine set on RINSE. If the engine and compartment are still considerably greasy, repeat the procedure.

An engine hoist or A-frame will also be necessary. Make sure the equipment is rated in excess of the combined rate of the engine and the accessories. Safety is of prime importance, considering the potential hazards involved in lifting the engine out of the vehicle. One of the more common methods of removing an engine is placing a chain hoist over a beam inside the garage, raising the vehicle and placing it on jackstands and then pulling the engine by lowering the vehicle. We don't recommend this method, and, unfortunately, when this method is used, the money saved by not buying or renting the proper equipment is often spent on repairing the garage roof or paying doctor bills! Another drawback of this method is that the vehicle must be moved before the engine will come out. It is much more difficult to move an automobile around an engine, rather than move the engine around inside the automobile.

Another common home mechanic method is the "A-frame" approach. Set up an A-frame made from three 11-to-15 foot-long, 5-to-6 inch diameter heavy steel poles set up like a tripod. Chain them securely at the top and hang a chain hoist from the chain. Drive your vehicle under the tripod and raise it off the ground and either support the front end on jackstands or use drive-up ramps. Be sure the ground is solid and will provide a safe area to work. Block the rear wheels to prevent the vehicle from rolling back. An A-frame properly built and secured is a lot stronger than a garage beam.

The strongest and most convenient device used to remove the engine, and the one we recommend, is the engine hoist or "cherry picker." One of these can be rented from your local equipment rental yard at a small fee plus a deposit. One type of cherry picker can be towed behind your vehicle and another type can be disassembled into eight or ten pieces and easily transported in a large vehicle.

Be sure to have the proper tools, drain pans, jackstands and other necessary equipment on hand. Two important items are a floorjack and jackstands. Be sure they are good quality and they are working properly to avoid any trouble when the vehicle is raised up. Have several drain pans handy so that the fluids can be separated and recycled into hazardous waste barrels. Most recyclers do not allow the coolant to mix with the oil or vice-versa. Check with the local Bureau of Automotive Repair, garbage disposal company or environmental authority to find out the locations of the pick-up sites. Also, other important items to keep on hand are fender protectors, wood blocks, cleaning solvent (to mop up any spills), masking tape (to identify vacuum hoses, electrical connections, etc.) and shop rags or towels.

Plan for the vehicle to be out of use for quite a while. A machine shop will be required to perform some of the work the do-it-yourselfer can't accomplish without special equipment. They often have a busy schedule, so it would be a good idea to consult them before removing the engine in order to accurately estimate the amount of time required to rebuild or repair components that may need work.

Always be extremely careful when removing and installing the engine. Serious injury can result from careless actions. Plan ahead, take your time and a job of this nature, although major, can be accomplished success- fully.

4.1 Use a permanent marker or scribe to mark the area around the edge of the hinge and bolt to indicate hood alignment

Removing the engine

Because this book covers so many different models, engines and years, the engine removal procedure will be general, but it will apply to all engines. Certain details will have to be performed by the home mechanic that are above and beyond the available information. The vehicle pictured in this book is a 1973 Chevrolet Nova with a 350 cu. in. engine.

First, disconnect the battery and remove it from the engine compartment. Be sure the cell caps (if equipped) are in place, since acid could splash up into your face when the battery is set down.

Place fender covers over the fenders. Use blankets or something similar if professional type fender covers are not available.

Next, remove the hood from the vehicle. Most cars and trucks are equipped with bolted hood hinges. Before loosening the bolts, mark the bolt locations on the hood hinges with a permanent marker or scribe **(see illustration)**. This will insure correct alignment when the hood is installed. Don't rely on your memory! A little extra time spent marking the location of hinges, body parts or electrical parts can save hours of time when the engine is being installed later on.

Because of the weight and the awkward size, the hood must be removed by two people. Have your assistant hold the hood in the open position while the bolts are being removed. Place rags under the corners of the hood to prevent the hood from moving suddenly and scratching or denting the body. Carefully lift the hood and carry it over to a safe storage place.

Use masking tape to mark any electrical wires and vacuum lines and then disconnect the hood light accessories and the vacuum or water lines (windshield washer accessories).

Choose a place to store the hood in the shop or garage

4.2 Place the jackstands under the frame of the vehicle and set them both at a uniform height

4.3 Remove the sheet metal screws (arrows) from the fan shroud

that is safe and out-of-the-way. The spot must be far enough away from the work area to avoid the possibility of a tool falling on the surface and causing damage. If the hood is stored on end, tie the hood latch to a nail or plaster anchor with a piece of heavy-gauge wire.

When working on a van, the front hood, grille and miscellaneous body parts must be removed in order to remove the engine from the front of the vehicle. Refer to your particular vehicle's *Haynes Automotive Repair Manual* for more details. Carefully look over the van to determine the easiest method of engine removal based on the height of the van, the position and weight of the engine.

Raise the vehicle and secure the chassis with jackstands. Be sure the area you have chosen is the right one, as the vehicle will be out of commission for a while and consequently will not be moved unless there are unusual circumstances. When jacking up the vehicle, place the jack under the crossmember that spans the engine (if equipped). This will balance the front end when it is being raised. Some Uni-body cars are not equipped with this crossmember, so use an alternate spot that is solid (e.g. the front suspension strut-rod bracket) and raise the vehicle. Consult your vehicle's owner's manual for the locations of the jacking points. Do not attempt to place the floorjack on any of the brackets that support the exhaust system! Place the jackstands in a uniform pattern set to the same height **(see illustration)**. It is only necessary to raise the front of the vehicle to remove the engine, but if you prefer working on a vehicle that is level, place jackstands under the rear end. Be sure they are set at the same height also.

Now, walk around and survey the vehicle and give it a nudge or two to test just how solid it is sitting on the jackstands. If there are any doubts about your safety, reposition the jackstands or find a more suitable place to work, but be satisfied the vehicle is well secured before continuing with the job.

Drain the fluids from the engine, cooling system and automatic transmission (if equipped). Set a pan (large enough to hold two or three gallons of fluid) under the radi-

ator drain plug and open the plug. Remove the radiator cap to bleed off any air pressure and to allow the fluid to drain faster. If the radiator does not have a drain plug, remove the lower hose clamp and hose and let it drain from the inlet pipe. Be careful when releasing the coolant using this method: the coolant flows rapidly and in a large volume. Using a separate container or drain pan, unscrew the oil drain plug from the oil pan and let the oil out of the engine. It is a good idea to remove the oil filter and pour the old oil out into the drain pan as well and then reinstall the oil filter temporarily. This will protect the threads on the oil filter housing when the engine is lifted out. Disconnect the transmission cooling lines (if equipped) from the radiator and the transmission, and let the fluid drain into a separate container or pan. If the transmission fluid was recently changed and is relatively clean, make sure the pan is clean so the fluid can be reused.

Removing the Radiator

The first parts to be removed from the engine compartment are the radiator and its related components, which should be disconnected first. The radiator fins are easily bent and damaged, making it virtually impossible to remove the engine with the radiator in the vehicle. The core fins on the radiator are very sharp, so watch your knuckles when you are working nearby. Make sure all the coolant has drained from the system and the pan has been removed and the coolant properly disposed of (many communities have collection centers that will dispose of oil and coolant properly). You will have to "roll around" on your creeper under the vehicle to disconnect clamps, splash pans, transmission cooling lines etc., so it is a good idea to make sure the area is clean and there are not any obstacles.

Remove the bolts or sheet metal screws that hold the fan shroud onto the radiator **(see illustration)**. Some fan shrouds use tabs to keep the bottom locked into the frame of the radiator or body. Other shrouds are bolted or screwed into the radiator at the bottom. Use a drop light or flashlight in the dark corners of the engine compartment to

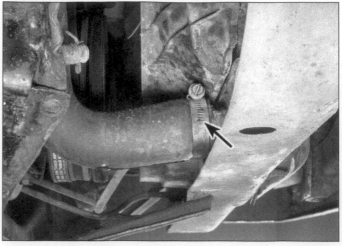

4.4 Disconnect the transmission cooling line fittings (arrows) from the radiator - use a back-up wrench on the radiator side of the fitting (if equipped) while unscrewing the larger outer nut on the fitting

4.5 Loosen the clamps (arrows) and remove the heater hoses from the engine

help identify the correct size of the bolt to avoid making a mistake (e.g. using an oversize socket or wrench and stripping the head). After disconnecting the fan shroud, set it away from the radiator, directly over the fan.

If the vehicle is equipped with an automatic transmission, disconnect the transmission cooling lines from the radiator and the transmission **(see illustration)**. Use a flare-nut wrench to avoid stripping the head of the fitting. If the fitting is very tight, do not force the fitting loose because there is a good chance the radiator will break at the junction. Use a back-up wrench on the radiator union to keep it stationary. Remove any clips or brackets that hold the transmission lines to the engine and carefully disconnect the lines. Try not to bend the lines too much when disconnecting them.

Disconnect and remove the radiator clamps and hoses from the radiator and engine block. If the hose is unusually tight around the connections, twist the hose with your hand until it is free. If they still won't budge, wrap the stubborn

4.6 Set the fan shroud back and remove the radiator

end with a shop rag and, using a large pair of adjustable pliers, squeeze and then twist the hose with leverage. Usually a hose that is corroded will have to be replaced with a new one. It is a good idea to replace all the radiator hoses if they are more than two years old or if the vehicle drives frequently in hot, dry climates. Disconnect the heater hoses from the engine block or the heater core (usually on the passenger's side of the firewall) **(see illustration)**. Watch out for excess coolant that did not drain from the heater core!

Remove the bolts that hold down the radiator. Some types of radiators (crossflow) are secured to the frame at the top of the radiator with a long bracket. This type is usually held in place at the bottom with tabs that are inserted over rubber brackets. Other types of radiators (downflow) are bolted to the frame on the top and bottom. Remove all the bolts and brackets from the radiator and lift the radiator out **(see illustration)**. Be careful not to damage the cooling fins by knocking it against the fan or the frame of the vehicle. Find a safe spot in the shop or garage and lay the radiator down flat.

Bring the radiator to a professional radiator repair shop and have it tested for leaks, cooling capacity and general condition. At best, the radiator will just need a flush and a checkover. If the core is blocked but salvageable, the radiator will require a rod-out (soak in a anti-corrosive chemical solvent). If the repairman deter- mines the radiator core is permanently damaged, the radiator will need to be recored (frame removed and a new core soldered into place). The prices vary quite a bit, so hope for the best but prepare for the worst. At any rate, it is of extreme importance that the radiator is diagnosed in excellent working condition to insure that the rebuilt engine will stay cool under all conditions (e.g. hot days, under load, freeway driving etc.). A good working radiator is by far the cheapest form of insurance you can buy for your rebuilt engine.

Now that the radiator is out, remove the fan shroud from the engine compartment and then remove the fan. Use a box end wrench, preferably one with a long handle) and

4.7 Use a box-end wrench to remove the fan assembly bolts

4.8 Use a flat blade screwdriver to pry the throttle cable from the carburetor or throttle body

loosen the bolts **(see illustration)**. Hold the fan to keep the pulley from moving. On clutch-drive fans, use an open-end wrench to reach the nuts or bolts between the fan and pulley. If the fan is a viscous-drive or clutch type, store it face down to prevent the fluid from leaking out.

Removing the air cleaner

Remove the air cleaner element and the air cleaner housing. Use masking tape and carefully mark each hose and electrical connector that is attached to the air cleaner housing. Disconnect all the vacuum lines, hoses and electrical connectors and remove the housing. Be careful so you don't damage any of the tubes or hoses.

Disconnecting the throttle cable linkage

Disconnect the throttle cable from the carburetor or fuel-injection unit. Most of these models are equipped with either the rod-and-lever type or the cable type. If it is the rod-and-lever type, disconnect it from the carburetor linkage and set if off to the side. If it is the cable type, disconnect the cable at the carburetor or fuel-injection unit and use needle-nose pliers to pinch the tabs and pull the cable out of the bracket on the intake manifold **(see illustration)**. If you have to remove any clips from the rods, pop them back on after the throttle cable or linkage has been disconnected. That way you won't have to worry about losing them! If the vehicle is equipped with an automatic transmission, disconnect the throttle valve (TV) cable or kickdown rod from the carburetor/fuel injection unit. Wire it to the firewall or somewhere out of the way.

Removing accessories

Warning: *Gasoline is extremely flammable, so take extra precautions when you work on any part of the fuel system. Don't smoke or allow open flames or bare light bulbs near the work area, and don't work in a garage where a natural*

gas-type appliance (such as a water heater or clothes dryer) with a pilot light is present. If you spill any fuel on your skin, rinse it off immediately with soap and water. When you perform any kind of work on the fuel system, wear safety glasses and have a Class B type fire extinguisher on hand.

Before disconnecting any accessory item, it is a good idea to obtain some masking tape and carefully mark each electrical connector, vacuum hose or fuel line or anything that has to be disconnected in order to remove the engine. Place one piece of tape on each side **(see illustration)** and clearly label each piece of tape so that it will be obvious where each connection belongs when the engine is ready to be "hooked up." For example, on a carburetor solenoid valve, label each side CARB.SOL. or on the air conditioning compressor, label it A/C COMP. Don't be fooled. What is easy to disconnect NOW is confusing and time consuming to reconnect LATER!

Before removing any fuel lines, depressurize the fuel system. On carbureted vehicles, simply wrap a fuel line connector with a few shop rags and carefully back-off the

4.9 Use masking tape to mark each end of the various electrical connectors, hoses and vacuum lines before disconnecting them

4.10 Loosen the pivot bolt (A) and remove the adjustment bolt (B) to remove this air conditioning belt

4.11 Remove the bolts (arrows) that hold the air conditioning bracket to the engine

bolt and let the excess fuel drain into the rags. Sometimes, if the lines are difficult to loosen, use back-up wrenches to support the fuel line and prevent it from twisting. On fuel-injected engines, consult the owner's manual and disable the fuel pump by removing the fuse that governs the fuel pump and relay. This will allow the pressure to dissipate. For detailed information on depressurizing fuel-injection systems, follow the procedure in the *Haynes Automotive Repair Manual* for your particular vehicle.

Now, disconnect the electrical connectors, vacuum lines, hoses, fuel injection components or anything that is connected to the engine. Disconnect the harness wires from the oil pressure sending unit, water temperature sending unit, distributor and coil, emission control units (EGR system, EFE system etc.), warm air hoses or valves and carburetor devices. Do not force any of the clips on the electrical connectors, instead use a small, narrow screwdriver and press the release tabs.

Loosen the belts and remove the air conditioning compressor (if equipped) from the engine. It is better to have the air conditioning compressor left attached to the air conditioning system than to have the compressor discharged and separated. If you choose to disconnect the A/C compressor lines, the system must be discharged by a professional A/C shop and when the engine is running again, the system must be recharged with freon. All this labor exchange will add a considerable amount to the bill. **Note:** *If you disconnect the lines, be sure to plug all disconnected lines and fittings to prevent entry of moisture and dirt - the two worst enemies of an air conditioning system!* **Warning:** *The air conditioning system is under high pressure. DO NOT disassemble any part of the system (hoses, compressor, line fittings, etc.) until after the system has been depressurized by a dealership service department or service station.*

Release the tension on the belt(s) **(see illustration)**. Loosen both the pivot bolt and the adjustment bolt or nut. Remove the belt(s). If the belt cannot be removed because of its position on the main pulley, lay it to the side - out of the way.

Remove the air conditioning compressor from the engine. Depending on the type of A/C compressor and the mounting bracket(s), remove the bolts that attach it to the cylinder head and the block. If the compressor bolts are impossible to reach, remove the bolts that retain the compressor bracket(s) to the engine **(see illustration)** and remove the compressor with the bracket. Use some strong mechanic's wire and tie it to the engine compartment firewall.

Remove the bolts that hold the power steering pump to the engine **(see illustration)**. Here again, it is better to keep the power steering hoses attached to the pump to avoid a lot of extra work. After the power steering pump has been removed, use heavy-gauge wire or mechanic's wire and tie it against the engine compartment firewall. Try to tie it off as far as possible from the engine to allow plenty of room for the engine to move around once it is loose from the transmission. Be sure the top of the pump remains up so no fluid will leak from the reservoir.

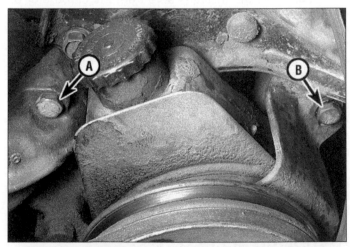

4.12 If your power steering pump bracket looks like this, loosen bolt (A), then loosen bolt (B) to remove the tension from the power steering belt - remove the bolts that hold the lower portion of the power steering pump bracket to the engine block (not in photo)

4.13 Remove the two long bolts (arrows) from the starter motor

4.14 Remove the bolts (arrows) and cover plates to gain access to the torque converter nuts

Remove the starter motor from the underside of the engine **(see illustration)**. Disconnect the wires from the starter (label them) and position it out of the way. Remove the bolts that hold the starter; the top bolt may be difficult to reach, so use an extension and a socket to reach inside the tight space. Lower the starter. Watch out! The starter looks small but it weighs quite a bit.

Check over your work for correct labeling, make sure the parts removed are in order, all the accessories are removed and the area around the vehicle is clean before continuing with the removal.

Disconnecting the torque converter

If the vehicle is equipped with an automatic transmission, remove the torque converter nuts. First, remove the front cover plate bolts **(see illustration)** and lower the plate. Be sure to mark the bolts or file them in a box that is labeled. The bellhousing bolts are odd sizes, so beware of the differences to help you when you install the engine. Before removing the torque converter bolts, use white paint and mark the torque converter in relation to the driveplate

(see illustration). This will insure that the driveplate and torque converter are assembled in the exact same alignment pattern. Also, it is a good idea to use a six point socket on the torque converter nuts to prevent any rounding-off of the nut head. Removing stripped and damaged torque converter nuts is an unnecessary waste of time.

Rotate the torque converter by turning the crankshaft. Use a deep socket and a 1/2 inch drive ratchet to turn the crankshaft pulley. Watch carefully for each torque converter nut (there are normally three) as you rotate the pulley. It is easier to remove each nut in order and prevent the time-consuming job of searching for the one torque converter nut you missed! If the engine you are removing has a frozen crankshaft (seized rod or bearing), you will have to remove the torque converter with the engine. Be very careful not to damage the input shaft or the torque converter.

Disconnecting the exhaust system

Disconnect the exhaust pipes from the exhaust manifolds **(see illustration)**. Before trying to remove those old, corroded flange nuts, spray them with a reliable penetrating

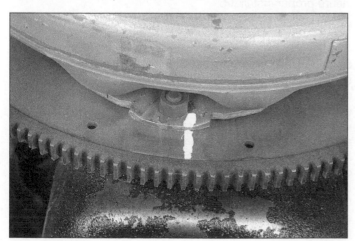

4.15 Use white paint to mark the torque converter in relation to the driveplate

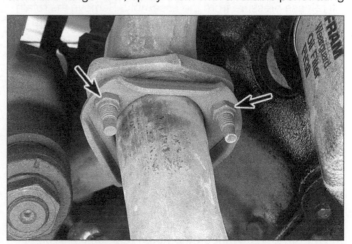

4.16 Remove the nuts (arrows) the hold the exhaust pipe flange to the exhaust manifold (third bolt hidden from view)

4.17 The chain must be connected to the engine with the proper amount of slack to avoid breaking the carburetor or other components when tension is applied with the engine hoist

4.18 Use a special tool to adjust the angle of the engine while it is being lifted from the engine compartment

lubricant. Allow several minutes for the liquid to soak into the threads. Use a six-point deep socket to avoid rounding-off the nuts or bolts.

Use an extension on the ratchet, but try to keep the extension as short as possible to avoid losing leverage. Disconnect any brackets or clamps on the exhaust system to allow the pipes to drop down low enough for ample clearance for removing the engine.

Disconnect the clutch cable or the clutch linkage

Disconnect the clutch cable or the clutch linkage from the engine and transmission. Depending on the type of system that is installed on the vehicle, make sure the system does not interfere with the removal of the engine. On the rod-and-lever clutch linkage, disconnect the return spring which is attached to the frame and the release lever. This will release pressure from the assembly. Remove the retaining clips and the rod from the linkage assembly. Replace any worn parts with new ones. Remove the cross-shaft from the engine and frame. It pivots on a bracket that is attached to the frame. After removing the bolts and bracket, the cross-shaft can be easily removed from the pivot assembly on the engine.

Detaching the engine from the vehicle

On hydraulic release systems, it may be necessary to unbolt the release cylinder. Do not disconnect the hydraulic line or you'll have to bleed the clutch release system.

Disconnect all those "little" things that you have been putting off. Don't overlook the automatic transmission filler tube and the engine ground straps. The filler tube is usually attached to the rear of the cylinder head, but, unfortunately,

it is tough to reach. Use a long box-end wrench and slowly work the bolt out. Lift the filler tube out by twisting and lifting to allow the bends in the tube to clear the bellhousing and engine. The ground straps are easier to remove but hard to find. On some models they are attached to the bellhousing near the bottom and others are attached to the engine cylinder head. Just look for a thick, braided wire shaped like a strap, usually with no insulation. Look over the engine and double-check that everything is disconnected. It is best to check everything on the top of the engine first, then roll under the vehicle with your creeper and systematically check all the connections underneath.

Unbolt the transmission bellhousing from the engine block. The upper bolts are very close to the body, so it might be necessary to use either swivel sockets on the end of long extensions or swivel joints linked between extensions to "bend" around corners. The choice of tools is very important for this step. Sometimes, experimenting with different arrangements is in order. Keep in mind that the bellhousing bolts are different lengths; mark them carefully to insure proper installation.

Installing the engine hoist

Use a heavy-duty chain and attach it to the engine. Some engines are equipped with lifting lugs or hooks. They usually are connected to the cylinder head with the exhaust manifold(s) and bolts. Sometimes they are located at the ends of the intake manifold. Use a heavy-duty bolt, nut and washer and tighten chain to the lifting lugs. Route the chain across the engine, leaving enough chain slack to allow the hoist to lift the engine without damaging any components once the chain is tight **(see illustration)**. On the other hand, the chain cannot be too long or the engine will not clear the front of the vehicle when it is being lifted out. Roll the engine hoist or "cherry picker" over the vehicle and attach the hook onto the middle of the chain. This will allow the engine to balance once it is detached from the vehicle. If the engine is not equipped with lifting lugs, go to an auto

4.19 A small-block engine mount

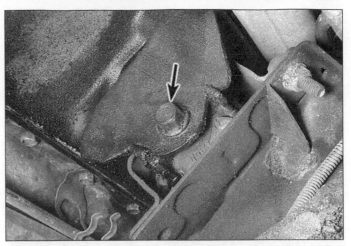

4.20 Remove the through bolt (arrow) and remove the engine with the upper mount attached to the block

parts store and buy some or wrap the chain through the exhaust manifold and bolt the ends to the chain. **Note:** *If the chain links are too large to fit through the exhaust manifold, it will be necessary to obtain the lifting lugs. Other ways may not be safe and reliable!*

On vehicles with small engine compartments, the engine might have to be tilted slightly in order to clear the front end of the engine compartment. Use a special tool that allows adjustment of the engine angle as it is being lifted from the engine compartment **(see illustration).** The tool is available at auto parts stores or automotive performance shops. If one is not available, estimate the amount of working area you have and set the hook onto the chain two or three links back from center to give a slight angle after the engine has been raised. Raise the engine hoist until the chain is tight and the engine first starts to lift.

Disconnecting the engine mounts

Disconnect the engine mounts from the engine block **(see illustration).** Engine mounts come in many sizes and shapes, depending on the size of the engine and the type of

vehicle. Basically, the engine mounts are separated into upper and lower sections and they are joined by a through bolt **(see illustration).**

Raise the engine slightly. Once the weight is removed from the engine mount, remove the through bolts. There are certain models that will require the engine mounts to be removed entirely before the engine will lift out.

Pulling the engine from the vehicle

Roll the floor jack under the transmission and secure the transmission so that it will not drop suddenly when the engine is removed. Be sure to place a piece of wood between the transmission and the surface of the floor jack. Raise the engine partially out of the engine compartment and separate the engine from the bellhousing with a long screwdriver or prybar wedged between the two **(see illustration).** Pull the engine as far forward as possible **(see illustration)** and raise it again until the engine is clear of the engine mounts. Keep the jack contacted to the transmission. Double-check that all accessories, wires, hoses etc.

4.21 Use a prybar or a large screwdriver and pry the engine from the transmission bellhousing

4.22 Gently rock the engine, pulling forward while you raise it slightly - do not force the engine and watch carefully for anything that has not been disconnected

4.23 Do not scratch the body or bend the emblem as the engine crosses the top - if you can't raise the engine high enough, you may be able to have assistants sit on the front of the vehicle to lower it

4.24 Use a piece of pipe to straddle the engine compartment - install long bolts into the transmission to support it

are properly disconnected and raise the engine out of the vehicle **(see illustration)**. Do not force the engine out of the vehicle; if it hangs up somewhere STOP and look over the situation carefully and determine exactly what is causing the problem. It usually is something that was overlooked such as a temperature sensor wire, ground strap, tie-down or clamp, etc. Make the necessary disconnection's and continue with the removal.

Raise the engine over the front section of the vehicle and roll the hoist back until the engine is away from the vehicle. Lower the engine and set it on a wooden pallet or onto the floor, if the hoist will extend that low. Set the engine down until the weight is off the engine, but do not disconnect the hoist from the engine yet.

Install a piece of pipe across the frame to support the transmission **(see illustration)**. Use bolt(s) or heavy-gauge wire to hold the transmission to the pipe. Remove the floor-jack from under the transmission and temporarily set it in

the corner of the shop.

Remove the driveplate (automatic transmission) or the clutch assembly (standard transmission) from the engine.

On engines equipped with a driveplate, use white paint and mark the position of the driveplate to the crankshaft **(see illustration)**. This will insure correct alignment when it is installed later on. Remove the bolts from the driveplate. This is easiest with an air impact wrench. If you don't have one, slide a screwdriver through one of the holes in the driveplate to lock it in place while you loosen the bolts. Now is a good time to gather shoe boxes or jars to keep all the engine bolts, nuts and miscellaneous items organized.

On engines equipped with a clutch assembly, mark the position of the pressure plate and flywheel before removing the bolts. Remove the pressure plate and friction disc and put the bolts into a box that is labeled. Note which end of the friction disc faces the flywheel so you can install it the same way. Next, unbolt the flywheel. This is easiest with an air impact wrench. If you don't have one, wedge a large screwdriver between the flywheel ring gear teeth and a pro-

4.25 Use white paint to mark the position of the driveplate on the crankshaft

4.26 Use long high-strength bolts (arrows) to hold the engine block on the stand - make sure they are tight before lowering the engine and resting all the weight on the stand

trusion on the rear of the engine block to lock the flywheel in place as you unbolt it. Be very careful as you remove the last bolt and pull the flywheel off the engine - the flywheel is very heavy!

Now that the rear section of the engine block is exposed, raise the engine and install it on the engine stand. Set the engine stand inside the legs of the engine hoist and raise the engine until it is a few inches above the height of the stand. Remove the engine brace from the stand and install it onto the engine block. The brace has four adjustable flanges that can be positioned to bolt up to any size engine block. Check that all the spacers and bolts are uniform and the thread pitch is the correct size and tighten all the bolts. Place the engine stand over the brace **(see illustration)** and lower the assembly onto the floor. Do not completely set all the weight of the engine down at first, but instead check all the bolts and the engine stand to make sure it is steady and secure. Unhook the engine hoist and the chain from the engine and roll the engine hoist into the corner of the shop. Now you are ready to remove the components from the engine block.

Removing the external engine components

Look over the entire engine. The next job is to remove all the engine components from the engine block, clean all the parts, and to remember where and how each one is assembled back onto the block. Sounds easy. In truth, each component has a particular place and function, but unfortunately they are assembled like a puzzle. Check very carefully for any options that might be installed on your engine (e.g. cruise control, air conditioning, power steering, etc.) and make notes concerning their electrical connectors, brackets, hoses etc. Often, it helps to make sketches of how things are hooked up or take instant photographs. As good a memory as you may think you have, you're bound to forget at least a few details during the weeks the engine is disassembled. Trying to figure things out can be very time consuming and frustrating! Remove each component carefully and in the correct order and the assembly process will be simplified. Air tools are a big help when unbolting the components, but if you don't have them, a speed handle for your sockets will also speed the process. A speed handle (available from most tool manufacturers) is a long rod bent into a crank with a 3/8 or 1/2-inch drive at one end. It enables you to quickly "crank" out bolts for the valve covers, oil pan, water pump, etc.. When removing components, follow this basic order:

1 *Valve covers*
2 *Remaining accessories (Alternator, air pump, A/C compressor, power steering pump)*
3 *Exhaust manifolds*
4 *Carburetor or throttle body*
5 *Distributor*
6 *Fuel pump*

7 *Thermostat*
8 *Intake manifold*
9 *Water pump*
10 *Engine mounts*
11 *Vibration damper*
12 *Timing chain cover*
13 *Oil pan*
14 *Oil pressure sending unit*

Valve covers

The first items to be removed from the engine are the valve covers. Remove the spark plug wires and distributor cap from the engine - it's best to leave the wires attached to the cap and mark each spark plug wire according to the number of the cylinder. This will make the installation quick and simple. Otherwise, you may get the firing order mixed up and get caught figuring out why the engine backfires after spending so much time rebuilding it! The cylinder numbers are usually stamped on the intake manifold, above the spark plug.

4.27 Remove the bolts (arrows) that secure the valve cover to the cylinder head - 350 engine shown

Refer to the your particular vehicle's *Haynes Automotive Repair Manual* for any additional details. Remove the bolts that retain the valve covers to the cylinder heads **(see illustration)**. Remove the valve covers and scrape any gasket material from the flanged surface with a gasket scraper. Thoroughly clean the inside and the out- side of the valve cover(s) with a parts brush in a solvent tank. If a solvent tank is not available, save the valve covers in a box with other greasy external engine components and bring them to a self-serve car wash. High-pressure hot water removes grease and dirt fairly well.

Remaining accessories

Warning: *The air conditioning system is under high pressure. DO NOT disassemble any part of the system (hoses, compressor, line fittings, etc.) until after the system has been depressurized by a dealership service department or service station).*

If all the accessory components are assembled onto

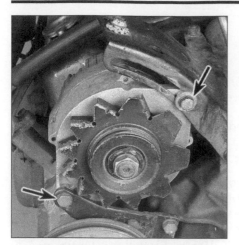

4.28 Remove the alternator bolts (arrows) from the brackets

4.29 Remove the bolts (arrows) and the pulley from the air pump

4.30 Remove the pivot bolt (arrow) and the bolts that retain the bracket to the engine block

the block, it is easier to remove the alternator and the air pump first. The air conditioning (A/C) compressor and power steering pump should be removed from the engine before the engine is removed from the vehicle to not allow the system pressure and fluids to be altered. If you choose to disconnect the hoses from the A/C and power steering systems, the compressor and power steering pump may be removed later. Make sure all the drivebelts have been removed.

Remove the alternator bolts and brackets (see illustration). The alternator is attached to the cylinder head and the water pump with brackets and spacers. Organize the bolts and label each one to insure correct assembly. Make notes or even draw a diagram or take an instant photograph to show where each spacer and bracket is located. Here again, the bracket looks simple enough until it is removed and sits in a parts box weeks. It's a good idea to keep the alternator brackets separate from the other brackets to avoid confusion later on.

Remove the air pump and brackets from the engine block. First, loosen the adjusting bolt from the bracket and swing the pump back so that the belt can be removed (see illustration). Then remove the pulley from the air pump. Next, remove the upper bracket (see illustration). Remove the bolts and brackets and separate the air pump from the block. Take a little extra time and make notes or draw a diagram of the bracket and pump location to insure correct installation.

If not already done, remove the air conditioning compressor from the engine block. Refer back to illustration 4.11 to get an idea of the bolt locations for the compressor and bracket. If your compressor is different, take a close look at the bracket and the bolts that hold it to the engine block. If the compressor cannot be removed from the bracket, remove it from the engine along with the bracket. It is important that the bolts are labeled and organized to prevent installing ones with different lengths into the wrong holes! After the compressor is removed, check the oil level inside the compressor and if it is low or excessively dirty, it is a good idea to go ahead and change it.

Remove the power steering pump from the engine block and carefully locate all the bolts and brackets that retain the power steering assembly to the block. If your pump is different, take notes concerning the bolt locations, brackets and length of the bolts to help later with the installation. Remove all the bolts and the bracket(s).

Exhaust manifolds

Remove the exhaust manifolds from the cylinder heads. Spray the bolts and nuts that hold the exhaust manifold and the heat shield with a reliable penetrating oil. If the exhaust manifold bolts are equipped with lock tabs, use a chisel and bend them back (see illustration). Remove the bolts from the exhaust manifold (see illustration). Some models are equipped with a spark plug shield that must be removed to gain access to the manifold bolts (see illustration). If your particular model has any extra emission equipment (EFE, TAC etc.), be sure to label and remove it from the exhaust manifolds. Remove the exhaust manifolds from the block. **Note:** *Be sure to carefully inspect the exhaust*

4.31 Use a chisel and bend the lock tabs back to allow room for a wrench or socket

4.32 Use a box-end wrench or socket to remove the bolts

4.33 Remove the bolt (arrow) that holds the heat shield to the block

manifold(s) for cracks **(see illustration)**.

Once the exhaust manifolds have been removed from the engine, it is a good idea to check for warpage. If you do not have the proper tools, take the manifolds to a reliable machine shop and have a machinist check them over. Use a precision straightedge on the cylinder head-side of the exhaust manifold. Try to insert a feeler gauge under the straightedge at different points on the surface of the manifold. If the measurements are excessive (over 0.003 to 0.006 in), have the exhaust manifolds resurfaced at a machine shop. The machine shop should have the exact specifications for your year and model.

Carburetor/throttle body

Warning: *Gasoline is extremely flammable, so take extra precautions when you work on any part of the fuel system. Don't smoke or allow open flames or bare light bulbs near the work area, and don't work in a garage where a natural gas-type appliance (such as a water heater or clothes dryer) with a pilot light is present. If you spill any fuel on your skin, rinse it off immediately with soap and water. When you perform any kind of work on the fuel system, wear safety*

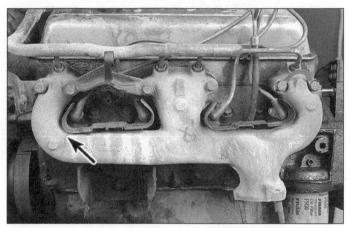

4.34 The tremendous amount of heat from the exhaust can easily crack the exhaust manifold (arrow)

glasses and have a Class B type fire extinguisher on hand.

Remove the carburetor or the throttle body from the intake manifold. First, disconnect all the hoses and vacuum lines and remove the front **(see illustration)** and rear **(see illustration)** mounting bolts from the base of the carbure-

4.35 Remove the front (small block engine shown) . . .

4.36 . . . and rear carburetor nuts or bolts (big block engine shown)

4.37 Turn the crankshaft until zero on the timing plate scale is directly opposite the notch on the vibration damper

4.38 Remove the distributor bolt and clamp from the base of the distributor - a special wrench, like the one shown here, is often helpful

tor/throttle body. Be sure to label everything properly before removing the carburetor/throttle body. Remove the unit from the manifold. Cover the top of the carburetor/throttle body with a clean rag to avoid any dirt or water from entering.

Distributor

On most models, it is necessary to remove the distributor before removing the intake manifold. It is a good idea to set the engine on Top Dead Center (TDC) before you remove the distributor. This will get you familiar with the correct alignment of the distributor rotor and the timing marks on it and timing plate. Use a 1/2-inch drive ratchet and a deep socket and turn the pulley over until the notch in the pulley is aligned with the 0 on the timing plate. If the timing plate is still greasy, spray some solvent on it and wipe it clean with a shop rag. In most cases, the number 0 indicates TDC, and it's surrounded by other numbers (e.g. 10, 20, 30 etc.) indicating the crankshaft degrees After Top Dead Center (ATDC) or Before Top Dead Center (BTDC) - depending on which side of 0 they're on **(see illustration)**. Remove the distributor cap and make a mark on the rim of the distributor with paint to indicate where the rotor is positioned. Check the spark plug wire on the corresponding terminal of the distributor cap and follow it to the number one cylinder. **Note:** *If the spark plug wires have been removed, look at the stamp on the wire or the label for the correct number. If the rotor is pointing directly opposite the number one terminal, rotate the crankshaft 180-degrees and realign the marks; the rotor should now be pointing to the number one terminal.*

Also, paint a mark on the base of the distributor and the intake manifold to ensure correct installation. This will allow you to insert the distributor with the vacuum advance unit in the exact same position to avoid interfering with other components.

Remove the distributor clamp **(see illustration)** and lift the distributor out of the intake manifold. If it is very stub-

4.39 If the distributor is stubborn, use a prybar and twist and lift the distributor out of the manifold - make sure you pry straight up and don't pry too hard; the cast aluminum distributor housing cracks easily

born, use a small crowbar or a large screwdriver and pry the distributor up **(see illustration)**. Do not force the distributor, but simply twist it while the upward force is applied.

Fuel pump

Warning: *Gasoline is extremely flammable, so take extra precautions when you work on any part of the fuel system. Don't smoke or allow open flames or bare light bulbs near the work area, and don't work in a garage where a natural gas-type appliance (such as a water heater or clothes dryer) with a pilot light is present. If you spill any fuel on your skin, rinse it off immediately with soap and water. When you perform any kind of work on the fuel system, wear safety glasses and have a Class B type fire extinguisher on hand.*

If the engine is fuel injected, leave the fuel pump mounted to the chassis, but, if the engine is carbureted and equipped with a mechanical fuel pump, remove it from the

4.40 The fuel line connections include a metal line with a hex fitting (A) and a rubber fuel line with a clamp (B)

4.41 Remove the bolts (arrows) from the fuel pump

4.42 Remove the bolts from the thermostat housing

4.43 Note the position of the thermostat in the manifold (the end with the arch faces out, as shown here) and remove it - if it's stuck, use a screwdriver and pry the thermostat out carefully

engine block **(see illustrations)**. Pour any left-over fuel trapped inside the pump into an approved gasoline container immediately after removing the fuel pump from the block.

On small-block engines, remove the plate under the pump (two more bolts) and slide out the fuel pump pushrod. On big-block engines, you'll have to remove the plug below the pump mounting surface before you can slide out the pushrod in the block. You may need a magnet to help pull it out.

Thermostat and thermostat housing

Remove the thermostat from the thermostat housing. First, remove any hoses connected to the coolant sensor in the housing (if equipped). Be very careful not to bend the fittings when pulling on the plastic parts. If it is stubborn,

use a small screwdriver and pry the hose out while simultaneously pulling. Remove the thermostat housing bolts **(see illustration)** and pull the housing off. Pull the thermostat out of the block **(see illustration)**. Use a new thermostat for the rebuilt engine unless the thermostat was recently changed. If that is the case, test it in a pan of water placed on a hot plate or stove. Raise the temperature and observe the temperature of the water with a thermometer. If the thermostat opens at the correct temperature (usually 160 to 185-degrees F), the part can be reinstalled into the rebuilt engine. Do not take a chance on a old thermostat without testing it.

Intake manifold

Remove the intake manifold from the engine block. Disconnect any hoses or lines that still might be connected.

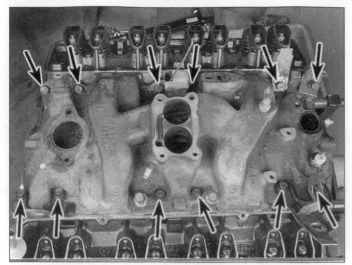

4.44 Remove the intake manifold bolts (arrows) -
350 engine shown

4.45 Remove the bolts (arrows) from the water pump

4.46 Remove the engine mount bolts (arrows) from the block

4.47 Remove the three pulley bolts (arrows) and lift the pulley
off the damper

Loosen the bolts from the intake manifold **(see illustration)** in several steps. Make absolutely sure you've located all the bolts, and note that some intake manifolds must be disassembled before removal. Label any bolts if they are not of the same length. This will insure correct installation. Lift the manifold from the engine block. If the intake manifold remains attached to the surface, use a small prybar or a large screwdriver and pry the manifold at a casting protrusion so you don't damage the gasket surface of the intake manifold and/or engine block. **Note:** *It's easy to not notice a hidden intake manifold bolt, particularly on a dirty engine. If the manifold does not come off relatively easily, double-check to make sure the bolts are all removed. Scrape any old gasket material from the surfaces and set the intake manifold aside to be hot tanked at the machine shop.*

Water pump

Remove the water pump from the engine block. First,

disconnect any hoses that still might be connected. Loosen and remove the bolts, noting the different diameters and lengths so they can be returned to the same holes **(see illustration)**. There may be empty holes from the bolts that were removed from the A/C bracket and the power steering pump Lift the water pump from the engine block. If the pump remains attached to the surface, use a small prybar or a large screwdriver and pry the water pump on a casting protrusion that will not damage the gasket surface of the water pump and/or timing chain cover. Scrape the old gasket material from the surfaces.

Engine mounts

Remove the engine mounts from the engine block **(see illustration)**. Label each side (RIGHT or LEFT). If the rubber insulation is hardened, cracked, deteriorated, separated from the steel backing or otherwise damaged, replace the unit with a new part. Since it's easy to do at this stage, we

4.48 Use a special tool (steering wheel and vibration damper puller) to remove the vibration damper

4.49 Remove the timing chain cover bolts (arrows) from the engine block (small block shown)

4.50 Remove the oil pan bolts and nuts - note the position and length of any different sized bolts

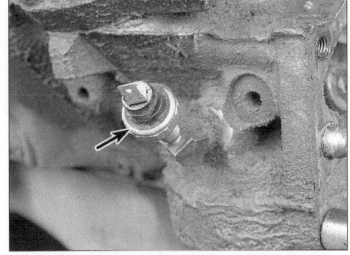

4.51 Use a deep socket and remove the oil pressure sending unit - the type shown here is for an indicator light; if you have a gauge on the dashboard, the sending unit will be larger and canister-shaped (big block engine shown) - On a small block, the sending unit is at the upper rear of the engine

recommend routinely replacing the engine mounts. When the engine is reinstalled, the mounts aren't nearly as easy to replace.

Vibration damper

Remove the small bolts and detach the crankshaft pulley **(see illustration)**, then remove the large bolt in the center of the vibration damper. An air impact wrench will make this job much easier; however, if one is not available, lock the engine in place, using one of the methods described in Pulling the engine from the vehicle above, then use a socket and large breaker bar to break loose the bolt - it's usually on very tight! Use a special puller **(see illustration)** and remove the vibration damper from the crankshaft. Watch carefully to make sure the pulley remains even with the timing chain cover as it is "backed-out." Any side-force will damage the crankshaft.

Timing chain cover

Remove the timing chain cover from the front of the engine **(see illustration)**. Label each bolt as to the length and its designated hole. This will make the installation easier. Use a small prybar to separate the timing chain cover from the engine block, if necessary. Do not damage the gasket surfaces. Pry at a casting protrusion that is not near the gasket surface! Remove the timing chain cover from the engine and remove the old gasket material from the cover. Clean it thoroughly in solvent or have it hot tanked at the machine shop.

Oil pan

Remove the oil pan from the engine block. Loosen all the bolts and remove them **(see illustration)**. If the bolts

Haynes Chevrolet engine overhaul manual

are different lengths or different diameters, label them so they can be returned to their original locations. If the oil pan does not separate from the engine block, use a rubber mallet or a hammer and block of wood to bump it loose. Remove the pan and clean the old gasket material from the block and the pan.

Oil pressure sending unit

Remove the oil pressure sending unit **(see illustration)** from the engine block. Use a socket designed for this purpose, an open-end wrench or a deep socket to avoid damaging the unit.

5 Overhauling the cylinder heads

Removing the cylinder heads

Note: *If this procedure is done with the engine in the vehicle, the intake manifold must be removed and the exhaust manifolds separated from the heads first. If the vehicle is* equipped with aluminum heads, see the Haynes Automotive Repair Manual for your particular vehicle for additional procedures and torque values.

1   Remove the rocker arm covers.
2   Remove the rocker arm nuts **(see illustration)**.
3   Lift off the rocker arms and pivots. As you remove each rocker arm/pivot set, wire the parts together **(see illustration)**. If you remove the parts in order (by cylinder number)

5.1 Remove the rocker arm nut (arrows)

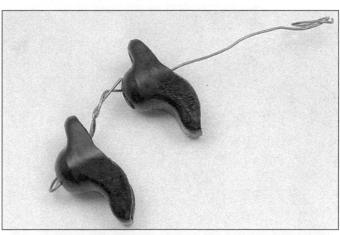

5.2 Wire each rocker arm/pivot set together, and string the sets sequentially, in cylinder order, on a continuous piece of wire - if the sets are wired in order, you only need to label the first set (cylinder No. 1)

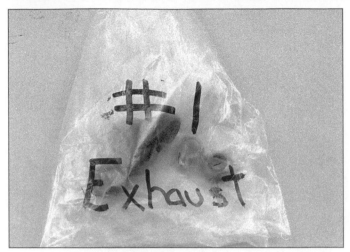

5.3 Label all components for each valve and place them in a separate bag

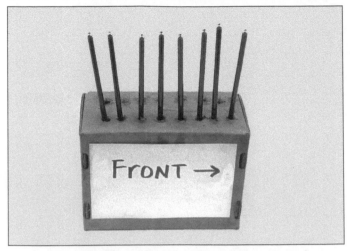

5.4 Use a cardboard box to make a pushrod organizer (drilling holes in a piece of wood will make a make permanent one)

and wire them to one long piece of wire, you only need to label the first set. You can also put them in plastic bags (label each bag if you do) **(see illustration)**. Don't mix up the parts - they should be returned to their original locations.

4   Lift out the pushrods and store them in a marked container so they can be returned to their original locations. If you're planning to overhaul cylinder heads on a regular basis, make a pushrod holder. Get a short section of 4 X 4 lumber and drill two rows of eight holes most of the way through. Label one end of the block "front" and store the pushrods in the holder, in order from front to rear, as you remove them. If you're in a hurry, or don't plan to make a habit of rebuilding heads, a cardboard box is adequate **(see illustration)**. Remember, don't mix up the pushrods if you're planning to reuse them. If you're installing new ones, it doesn't make any difference.

Each head is attached to the block with several bolts **(see illustration)**. The lower ones are evenly spaced along

the bottom of the head and the others are located near the valve springs. Remove all the bolts! If you forget to remove a bolt and try to pry off the head, you won't get it off and you may damage it.

6   To prevent the head from falling off the engine, run a bolt back into each end of the head just far enough to catch a couple of threads.

7   Insert a breaker bar into one of the intake ports and lever off the head. If that doesn't break the bond between the block and head, carefully insert a pry bar or sharp chisel between the block and head and dislodge the head **(see illustration)**. If you have to resort to this second method, make sure you don't damage the edge of the mating surface between the block and head.

8   Once the bond between the block and head is broken, remove the bolts and lift off the head. Label it "left" or "right."

9   Repeat Steps 1 through 8 for the other head.

5.5 Each cylinder head is attached to the block with several bolts (arrows) - working from the ends, remove them in a criss-cross fashion

5.6 If the head is stuck to the block, carefully pry it up at a casting protrusion (don't pry between the gasket surfaces!)

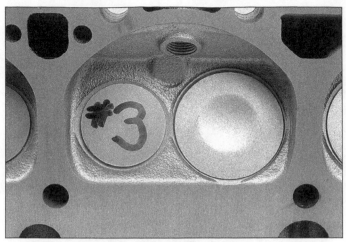

5.7 Another helpful strategy is to label each valve with a felt-tip pen

5.8 Install a C-type valve spring compressor over the head and compress the spring . . .

Rebuilding the cylinder heads

Rebuilding the heads is a dirty and tedious job, but it's got to be done right. Once you begin, try to focus on this one task until it's done, to reduce the likelihood of lost or mixed-up parts.

First, a word of advice: If you've never rebuilt cylinder heads, you're better off limiting yourself to removing and installing them. You can buy new and rebuilt cylinder heads for most Chevy engines at dealerships and auto parts stores. Swapping heads may be more practical and economical than disassembling, inspecting, reconditioning and reassembling the originals. To succeed in the head rebuilding game, you need some special tools and the skill to use them. Generally, this skill develops as you gain experience. You also need the right attitude. Head work is exacting and precise - if you approach it in a sloppy or disorganized fashion, you'll end up with junk.

On the other hand, if you've managed to round up the necessary equipment, and you're mechanically inclined -

then by all means, read on! We'll show you how to strip, clean and inspect the heads. This will give you a good idea of what needs to be done to restore the heads to like-new condition. But leave the precision work to a skilled automotive machinist. He probably won't even charge you any more to strip the heads.

Disassembling the cylinder heads

1   Fabricate a valve organizer from a piece of cardboard or a narrow strip of wood. Punch holes in the cardboard or drill holes in the wood to accept the valves as they are removed. As insurance, label each valve with a felt-tip pen (see illustration).

2   You'll need a valve spring compressor for this procedure (see Chapter 2). Install the compressor on the first valve in the head (see illustration), compress the spring, remove the keepers (see illustration), release the spring, set the compressor aside and remove the retainer, the spring and the valve. Repeat this step for each valve. As you remove the valves, put them in the valve organizer. Put the other parts in a clearly labeled plastic bag (see illustration).

5.9 . . . then remove the keepers with a small magnet or needle-nose pliers

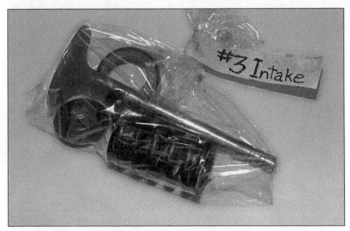

5.10 As you remove each set of keepers, valve retainer, spring, shims and valve, put them in a clearly labeled bag

5.11 If the keepers are stuck to their grooves in the valve stem, put a 3/8-inch drive, 9/16 inch deep socket squarely over the valve tip and, on the retainer, rap the socket sharply with a mallet - if the rap is sharp enough, it will pop the keepers loose and release the retainer and spring

5.12 If you can't pull the valve through the guide, deburr the edge of the stem tip with a flat file or a small whetstone - rotate the valve head while holding the file at an angle on the edge of the valve tip

3   It's not unusual for keepers to stick in the valve stem grooves. If any of the keepers are hard to remove, place the head right-side-up on the bench and put a small block of wood under the valve with the sticky keepers. Place a 9/16-inch deep socket squarely over the valve tip, on the retainer (**see illustration**). Rap the socket sharply with a soft-face hammer to loosen the keepers. If the blow is sharp enough, it will sometimes pop the keepers out and release the retainer and spring. It doesn't, reinstall the compressor and remove the keepers.

4   Pull the valve out of the head, then remove the oil seal from the guide. If the valve binds in the guide (won't pull through), there's a burr around the tip of the valve stem which is causing the valve to hang up in the guide. Push the valve back into the head and deburr the end of the valve

with a flat file or whetstone (**see illustration**). Rotate the valve head while holding the file at an angle on the edge of the valve tip.

Cleaning the cylinder heads

Note: *If you plan on having a professional rebuild the cylinder heads, skip the following cleaning and inspection procedures and take the heads directly to the machine shop now*. There they will be hot-tanked or baked in an oven (**see illustration**) and blown with compressed air (**see illustration**) to remove all deposits. They will also be carefully

5.13 Traditionally, heads have been "hot-tanked" to remove dirt and grime, but now machines like this heat the heads to about 750-degrees to clean

5.14 After the head is heated, it's hung out to cool off and the debris is blown off with compressed air

5.15 Use a scraper to remove all traces of old gasket material

5.16 Lightly polish the combustion chambers and intake and exhaust ports with a rotary brush

checked and all necessary new parts will be installed. When the heads are returned, they'll be ready to install on the engine with no additional work. If you choose to do the inspection procedures yourself, you will also have to do some preliminary cleaning, especially if the heads are extremely sludged or carboned up. After you're done inspecting them, you should still have them hot-tanked and bead-blasted by an automotive machine shop, even if you're going to do the service work yourself.

Warning: *Wear safety goggles when cleaning the heads. We also recommend wearing rubber gloves designed to prevent solvent from contacting your hands.*

1   Using a gasket scraper, remove all traces of the old gaskets from the cylinder head surfaces **(see Illustration)**. Most auto parts stores stock special gasket removal solvents that soften gaskets and make removal much easier. Remove all sludge, carbon deposits and oil with solvent and a wire brush. Decarbonizing chemicals can be helpful when cleaning cylinder heads and valve train components, but they're very caustic and should be used with caution. Be sure to follow the instructions on the container.

2   Wash the cylinder head with clean solvent and dry it with compressed air. **Warning:** *Wear eye protection when using compressed air!* Also blow out all coolant and oil passages.

3   Clean the rocker arms, nuts, fulcrums and pushrods with solvent and blow them dry. Don't mix them up during the cleaning process.

4   Clean all the valve springs, spring seats, keepers and retainers (or rotators) with solvent and dry them thoroughly. Do the components from one valve at a time to avoid mixing up the parts.

5   Scrape off any heavy deposits that may have formed on the valves, then use a motorized wire brush to remove deposits from the valve heads and stems. Again, make sure the valves don't get mixed up.

6   Using a stiff wire brush, remove any remaining deposits from the coolant passages and oil holes. Carefully scrape any residual carbon deposits from the combustion cham-

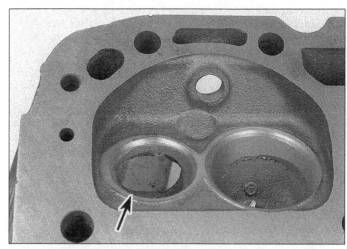

5.17 Inspect the head for cracks, particularly around the valve seats (arrow) - the head is usually junk if it's cracked, although your machine shop may be willing to "pin" it for you

bers and intake and exhaust ports, then clean them with a wire brush **(see illustration)**.

7   Clean the rocker arm stud threads with a wire brush. Run a tap into each threaded hole to remove corrosion and thread sealant. Blow all debris out of the holes. **Warning:** *Wear eye protection when using compressed air!*

Inspecting the cylinder heads

Checking for cracks

Inspect the cylinder head for cracks **(see illustration)**. Examine the valve seats in the combustion chambers very carefully. If they're burned, cracked or pitted, take the head(s) to an automotive machinist and see if repair is possible. If not, replace the head.

Checking cylinder head flatness

Cylinder heads are subjected to extreme heat and pressure. Because they're structurally weak compared to

5.18 Using a straightedge placed diagonally and across the head, measure the gap between the straightedge and the cylinder head -to -block gasket surface - if there's more than a 0.003 in gap between any two points within six inches of each other, or if the variation anywhere on the head exceeds 0.006 in., have the head milled

5.19 If more than 0.020 in. is milled from the heads, have the bottom and side of the intake manifold milled also so the intake ports, coolant passages and bolt holes will still be aligned

an engine block, they warp quite easily. If the engine has severely overheated, the heads are probably warped. Warping causes combustion and coolant leaks, so all gasket surfaces must be flat. Check the head gasket surfaces with a straightedge and feeler gauges. Set the straightedge lengthwise and diagonally across the head in both directions. Measure the gap between the head and the straightedge with a feeler gauge (see illustration). The maximum allowable variation in any gasket surface is 0.003 inch between any two points within six-inches. Over the length of the head, the variation should be no more than 0.006-inch for 10:1 compression ratio heads, or 0.007-inch for heads with a lower compression ratio. If you plan to use shim-type gaskets instead of the (more compliant) composition type, the acceptable warpage is half the above figures. If the head is not within specifications, it must be resurfaced (milled) by a properly equipped automotive machine shop.

Resurfacing (milling) the cylinder heads

If either head must be milled (by an automotive machine shop), have both heads done at the same time (see illustration). Otherwise, one head will have a slightly higher compression ratio than the other! The amount of material removed should not exceed 0.010-inch. Theoretically, you can remove up to 0.040-inch, but it's not a good idea if you're planning to use pump gas. This is particularly important if you're working on a pre-1972 high-compression engine. Milling the heads more than 0.010-inch on one of these engines is asking for trouble (detonation will occur).

If more than 0.020-inch is removed from the heads, the bottom and sides of the intake manifold must also be

machined to ensure the intake ports, coolant passages and bolt holes still line up. The machine shop should have a chart indicating how much to remove from the manifold.

Inspecting the rocker arm components

Examine the rocker arm studs. Look for damaged threads, broken studs and "notching". Notching occurs when a rocker arm rocks over on its side and the edge of the opening in the rocker arm contacts the stud, wearing a notch in the stud. Replace all notched studs; notching can lead to failure. Pressed in rocker arm studs should be replaced by an automotive machine shop.

Check the rocker arm faces (the areas that contact the pushrod ends and valve stems), the rocker arm pivot contact areas and the pivots for excessive wear, pitting, galling, scoring, cracks and rough spots.

Inspecting the pushrods

Check the pushrod ends for excessive wear. It's normal for the wear pattern at the rocker-arm end to extend farther around the ball than at the lifter end. But if the ball at the rocker-arm end is egg-shaped, replace the pushrod.

Roll each pushrod on a perfectly flat surface to determine if it's bent (see illustration). If it wobbles when it rolls, use a feeler gauge to measure the gap between the pushrod and the flat surface. If you can stick a feeler gauge thicker than 0.008-inch between the pushrod and the flat surface, replace the pushrod. Don't try to straighten it.

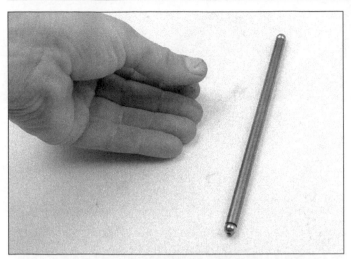

5.21 Roll each pushrod on a smooth surface, such as a plate of glass - if it wobbles when it rolls, check it with a feeler gauge - if you can get a gauge thicker than 0.008 in. between the pushrod and the surface, replace the pushrod

Inspecting the valve guides and stems

Note: *Valve guides are normally routinely replaced at overhaul time. If you find one questionable guide, it's a good idea to have them all replaced or machined.*

If the guides are badly worn, the valves won't seat squarely on the seats, allowing leaks past the valves, resulting in overheating and a loss of power. Worn guides also allow too much oil between the valve stem and the guide walls, resulting in high oil consumption, an oily exhaust and the buildup of an oily residue in the combustion chamber, which will foul plugs and raise the compression ratio, causing detonation.

5.22 Lay the head on a bench, pull each valve out about 1/8 inch, set up a dial indicator with the probe touching the valve stem, wiggle the valve and measure its movement

You can measure valve guide wear at the valve head or stem by using a dial indicator to measure the deflection of the valve when it's moved.

Lay the head on the bench. Set up a dial indicator perpendicular to the valve stem. Pull each valve out of the guide slightly (about 1/8 in.), then wiggle it as far as it will go in each direction **(see illustration)**. The total indicator reading must be divided by two to get the deflection, which should not exceed about 0.010 in.. In any case, the valve guides are usually replaced routinely at overhaul time.

For a more accurate reading, a hole gauge can be used to measure the size of the guide opening (close to the end), then a micrometer can be used to measure the valve stem diameter. Subtract the diameter from the opening measurement to get the clearance **(see illustrations)**.

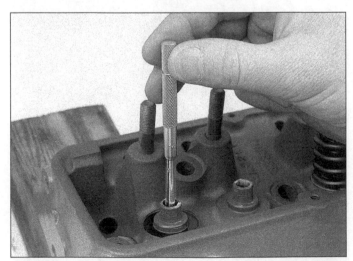

5.23a Or, for a more accurate reading, insert a small-hole gauge into the guide and expand it until it fits the guide with a light drag - check the guide bore at several places up and down the bore to locate maximum wear (usually near the bottom end)

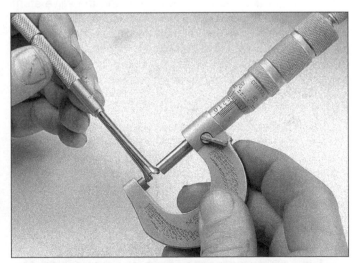

5.23b Each time you take a reading in the guide, pull the small-hole gauge out and mike it, then jot down your measurement on a diagram showing the location of each measurement - if the measurements vary much over 0.002 inch, the guide needs replacement or machining

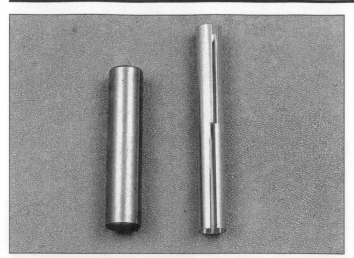

5.24 Replacement valve guides are made of cast iron, bronze or silicon-bronze and are available as "false-guides" (press-in replacements), thin -wall inserts and threaded inserts (not shown)

Reconditioning the valve guides

There are two common ways automotive machine shops recondition your valve guides: Knurling and guide inserts.

Knurling the guides

The knurling process displaces material and rolls a new, raised pattern into the guide walls. Knurling reduces the effective diameter of the guide, making it smaller in diameter than its previous size. Because the valve guide is now effectively smaller, it must be reamed to standard size. Reaming the guide smooths off the peaks - the high spots of the knurled pattern.

Knurling is inexpensive, but it's not cost effective. The actual contact area between the guide and the valve stem is substantially reduced, which causes a proportional increase in stem-to-guide contact pressure. The guide and stem therefore wear more quickly. So high oil consumption caused by worn guides won't be cured, it will simply be postponed. We don't recommend knurling.

Installing valve-guide inserts

Valve guide inserts **(see illustration)** restore the guides to their original - and sometimes even better - condition, depending on the material used. Cast iron inserts restore the guides to their original condition. To install them, the original guides are machined to the same inside diameter as the outside diameter of the new inserts, minus a few thousandths for an interference fit, then reamed to size, if necessary **(see illustration)**. This step isn't necessary for all inserts - it depends on the manufacturer.

Thin-wall bronze inserts offer better-than-original service life. There are two styles - the thread-in type and the press-in type. Press-in types, such as K-lines, are installed by reaming the existing guides **(see illustration)**, installing the inserts, trimming off the excess material and - using the

5.25 New cast-iron "false guides" are installed by machining the original guides to the same inside diameter as the outside diameter of the new guides - minus a few thousandths for an interference fit - pressing the new guides into place and reaming them to size (if necessary)

5.26 New silicon thin-wall inserts are installed by reaming the old guides oversize . . .

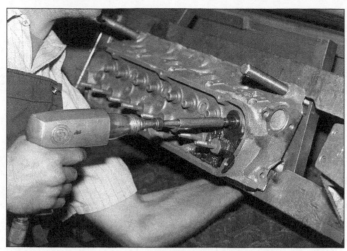

5.27 ... installing the inserts and "burnishing" (expanding) and reaming the insert

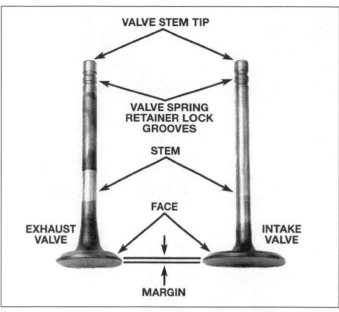

5.28a Inspect these areas on each valve

5.28b Look for pits and excessive war on the tip of the stem - this tip is a good example of what can happen if the guide becomes so worn that it allows the tip to flex under pressure from the rocker arm, or if a lifter seizes and the clearance between the tip and the rocker arm vanishes, or if the rocker nut or bolt loosens and allows the rocker to move around excessively

5.28c Inspect the valve face for signs of burning, uneven wear, deformation, cracks and pitting - signs of damage are usually a little more subtle than on this burned valve

original valve seat as a pilot to prevent tilting - "burnishing" (expanding) and reaming each guide to size **(see illustration)**.

The thread-in type of bronze insert is similar in appearance to a Heli-Coil except there are no inside threads. It's installed by threading the existing guide bore with the tap size specified by the manufacturer. A sharp tap must be used or the guide insert won't fully seat, which will allow it to gradually move up and down more and more and pump oil into the combustion chamber. The insert is threaded into the guide and expanded to lock it into place so it won't work sideways in the thread and so the backsides of the thread will be in total contact with the cast iron cylinder head for maximum heat transfer between the valve stem and the head. Finally, the guide is reamed to size.

Bronze and bronze-silicon guides are more expensive than cast-iron or all-bronze guides. But, because they're

easy on valve stems, they typically have a service life of 150,000 miles.

Inspecting the valves

Inspect the valves for obvious damage such as burned heads and excessively worn tips. Also look for signs of uneven wear, deformation, cracks and pitting **(see illustrations)**. Examine the valve stem for scuffing and galling and inspect the neck for cracks. Rotate the valve and check for any obvious indication that it's bent. Look for pits and excessive wear on the tip of the stem. If any of these conditions are present, have the valve(s) serviced by an automotive machine shop or replace them.

Compare the worn and unworn areas on each valve stem. Maximum stem wear usually occurs near the tip end of the valve - you'll recognize it easily because it's shinier than the rest of the stem and there's a sharp division

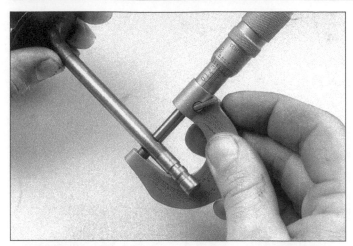

5.29 Measure the valve stem diameter with a micrometer

5.30 Valve faces are ground on a specially designed valve grinder: The valve is rotated in a collet-type chuck at an angle of 44 degrees to a high-speed grinding wheel; as the valve face rotates against the wheel's rotation, it's also oscillated across the face of the wheel; a cooling bath of cutting oil is poured onto the valve and grinding wheel during the operation to wash away the metal shavings

between it and the unworn area. The lower edge of the shiny area represents the maximum valve opening. This is the point at which the valve stops in its guide at its fully open position. Using a one-inch micrometer, measure the stem diameter of each valve immediately above and below the maximum-wear line **(see illustration)**, then subtract to determine wear.

How much valve-stem wear is acceptable? Well, it depends on how much service life you expect to get from your engine after the rebuild, how you reconditioned the guides and other factors. Check with your machine shop if you're unsure. **Note:** *The valves in pre-unleaded gas engines need the tetraethyl lead in leaded gas for lubrication.* If you're rebuilding an older engine, and you want to be able to run it on unleaded, get new "stellite-coated" valves designed to operate without lead. You'll also need to get new, hardened valve seat inserts specially manufactured for use with the new valves.

Reconditioning the valves

If your valves still have some life in them, have them reconditioned by a good automotive machine shop. Make sure they grind the faces and the tips. The face is the valve surface that contacts the valve seat in the head when the valve is closed. This seal must be perfect to prevent combustion gases from escaping into the intake and exhaust ports when the valve is closed by the valve spring.

Valve faces are ground on a purpose-built valve grinder **(see illustration)**. The valve is rotated in a collet-type chuck at an angle of 44 degrees to a high speed grinding wheel. As the valve face rotates against the wheel's rotation, it's also oscillated across the face of the wheel. A cooling bath of cutting oil poured onto the valve and grinding wheel washes away the grindings. Only enough material is ground off the face to expose new metal. If too much material is removed, the valve margin - the width of its outer edge at the face - will be too thin **(see illustration)**. A valve, especially an exhaust valve, with little or no margin is susceptible to overheating. As a rule of thumb, an exhaust valve should be at least 0.030 in., about 1/32 in., wide. An intake

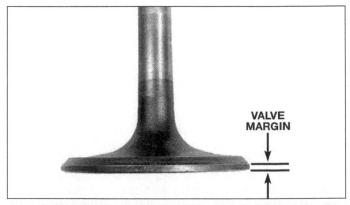

VALVE MARGIN

5.31 Only enough material should be removed from the valve face to expose new metal - if too much material is removed, the valve margin - the width of the valve's outer edge at the face - will be too thin and the valve will overheat

5.32 After the face is ground, the tip is faced - ground square to the centerline of the valve stem - using another attachment on the valve grinder

doesn't get as hot, so its margin can be a little narrower, but not less than 0.015 in., or about 1/64 in.

After the valve face is ground, the tip is faced (ground square to the centerline of the valve stem) with another attachment on the valve-grinding machine **(see illustration)**. Finally, the tip is chamfered (beveled) to eliminate the sharp edge created when the tip is faced. Your valves are now ready for installation.

Reconditioning the valve seats

The valve seats must also be reground by an automotive machine shop **(see illustration)**. Valve seat grinding equipment is expensive, highly specialized and takes a lot of skill and experience to operate. If the guides have been replaced, valve seat reconditioning is critical because the seats are no longer concentric with the new guides - they don't have the same centers. So even if a seat is in good condition, the valve won't seat against it properly when closed because the guide will position it off-center to the seat. A valve seat is ground with a mandrel centered in the valve guide, so once the seat is reground, it will be concentric with the guide.

Lapping the valves

Lapping the valves is a time-honored - but seldom used - tradition of grinding the valve face and its seat together with lapping compound. You apply lapping compound or paste to the valve face, install the valve in its guide and rotate the valve back and forth in a circular motion while pressing the valve head down against the seat. But, except for checking concentricity of the valve seat with the valve face, lapping is largely a waste of time. Why? Because if the face and seat aren't ground correctly, lapping usually won't make them right. And if they are done right, lapping shouldn't be necessary. The bottom line here is that there is no substitute for having the work done by a machinist who knows what he's doing.

5.33 The valve seats are ground using a machine like the one shown here - because the grinding stone is centered by a mandrel in the new valve guide, the seat is reground concentric to the new guide

Inspecting the valve springs

Note: *Valve springs are normally routinely replaced at overhaul time.*

Checking for squareness

Valve spring squareness is how straight a spring stands on a flat surface, or how much it tilts. Valve springs should be square so they load the spring retainers evenly around their full circumference. Uneven retainer loading increases stem and guide wear.

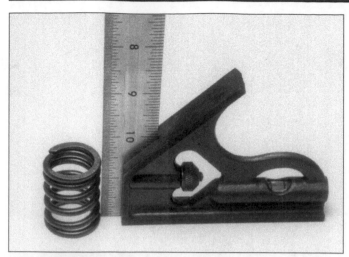

5.34 To check the valve springs, stand them up on a perfectly flat surface, place a square next to each spring and rotate the spring while keeping it up against the square - if there's a gap between the spring and the square, measure the gap - if the gap exceeds 1/16 in., replace the spring

5.35 A quick way to determine uniform spring height is to line them up on a flat surface and compare their relative heights - if the free heights of any of them vary by more than 0.625 in., replace them

For this check, you'll need a flat surface, and an accurate carpenter's square. Stand the spring up on the flat surface, place it against the square **(see illustration)**, rotate it against the square to determine its maximum tilt and measure the gap between the top of the spring and the vertical surface of the square. If the gap exceeds 1/16 in., replace the spring.

Checking spring free height

There's normally no way to check the spring load at installed height or at open height at home. A special spring tester is necessary for these two tests (see below). However, there are a couple of simple checks you can do at home to determine whether the springs are usable, and whether their condition merits further checking on a spring tester. First, check the spring free height.

Spring free height is the unloaded length of a spring. If a spring's free height is too long or too short, it's probably been fatigued or overheated by excessive engine operating temperatures. The load exerted by that spring when installed will be incorrect because it will be compressed more, or less, by its retainer than the other springs.

Line the springs up on a flat surface and compare their relative heights **(see illustration)**. If the free heights of any of them vary by more than 0.0625 in., they should be replaced. Here's a good rule-of-thumb guide: Replace any spring which is 1/8 in. shorter than its specified free height. It's okay to reuse a spring that's 1/16 in. or less shorter than its specified free height, but only in light service. Base your decision on the kind of use your engine will see and how much shorter the springs are. In other words, if the engine is going to be used for commuting to work and driving around town, you can use the marginal springs; if the engine is going to be driven hard, junk the marginal springs. They'll progressively weaken and the valves they control

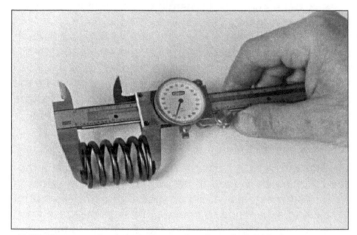

5.36 One way to determine spring free height is to measure each spring with a caliper

will eventually begin to "float," which will limit rpm. Your only alternative is to have the springs checked on a spring tester to determine whether they can be corrected by shimming.

If any of the spring free heights seem marginal, measure them more closely with a vernier caliper **(see illustration)** and compare your measurement(s) to the factory specifications (see the *Haynes Automotive Repair Manual* for your particular vehicle).

Checking spring load at installed height

A spring's load at its installed height is a common spring specification. A typical load-at-installed-height specification is 76 to 84 pounds at 1.820 inches. When compressed to a height of 1.820 inches from its free height, the load generated by the spring should be between 76 and 84 pounds. The absolute minimum installed load is 10 percent

5.37 You may be able to locate a tester like this that you can use with your bench vise to check spring load at installed height

5.38 If the engine has umbrella-type seals, wrap tape around the keeper grooves to protect the valve seal when it's installed - remove the tape after the seal is in place

5.39 Apply a small dab of grease to each keeper, as shown here, before installation - it will hold them in place on the valve stem as the spring is released

5.40a Make sure the O-ring seal under the retainer is seated in the lower groove and not twisted before installing the keepers

less than the lower load limit, or 68 pounds. If a spring doesn't exceed, or at least meet, the minimum standard limit, replace it. This test is normally performed by an automotive machine shop, but home testers may be available **(see illustration)**.

Sometimes, weak springs or springs with a too-short free height can be shimmed to restore their installed and open loads. To shim a weak valve spring, you place a shim - a special flat washer - between the spring and the cylinder head so the spring will be compressed more - thus increasing its load at open and installed height - when it's operated. With the shim installed, the spring load at its installed and open heights is rechecked on the spring tester. Excessive shimming can result in severe valve-train damage by causing the spring to coil bind (reach its solid height) before its valve is fully opened. Another problem is that a spring is designed to be compressed a certain amount. This amount is the difference between its free and open heights. If compressed more than this amount, it can be overstressed, become fatigued and "take a sack" (lose its load-producing capability).

Assembling the heads

Note: *Heads are usually assembled by the automotive machine shop performing the work on the heads.*

Once you've collected any new parts and all the old reconditioned parts from the machine shop, you're ready to reassemble the heads. You'll need your valve spring compressor and a six-inch steel ruler or vernier calipers. Be very careful when installing new umbrella-type valve seals over the valve stems. The keeper grooves can easily damage the seals, so use tape over the keeper grooves to protect the seals when installing them **(see illustration)**. Also, use a small dab of white grease on each keeper to hold it in place on the valve stem as the spring compressor is released **(see illustration)**. If you're working on a small block, equipped with O-ring seals, install the O-ring seal after the spring is compressed, being sure not to twist it and check it after installation **(see illustrations)**.

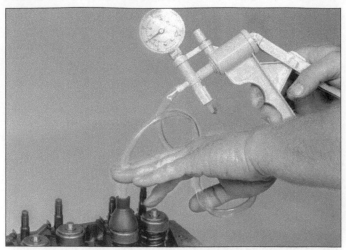

5.40b A special adapter and a vacuum pump are required to check the O-ring valve stem seals for leaks

5.41 The most accurate way to measure installed spring height is with vernier calipers - make your measurement is from the spring pad to the tip of the spring

Checking valve spring installed height

Every spring must be at its installed height, or its valve won't be loaded correctly. Remember installed height? That's the height the valve spring should be with the valve closed and the spring, valve, retainer, keepers and shim(s) (if installed) assembled (Shims are included in the spring's installed height IF they were used to correct installed height). If shims had to height, the shop performing the work should have already assembled the heads - with the shims in the correct places - for you.

Install the valves, seals, springs, retainers and keepers. The height is checked by placing a steel rule next to the assembled valve spring and measuring the height of the spring only (usually with a steel rule or vernier calipers) **(see illustration)**. Check the *Haynes Automotive Repair Manual* for your particular vehicle to get the correct specification. If the machine shop has done its work correctly, the height should be correct, but don't assume anything.

Checking the valve seal

After the heads have been assembled, verify that the reconditioned valves and seats are sealing properly before installing the heads. Place the heads upside down on the workbench with the head-gasket surface level. Fill each combustion chamber with kerosene and check in side the ports for leaks. If you don't see any, the valves are sealing properly. If you do see a leak, go back to the machine shop and have a little talk with your machinist.

Installing the heads

Position the block upright on the stand. Make sure the post is tight. We don't recommend trying to install the heads with block on the floor because the block is unstable as soon as you install one head. If you have to work on the

floor, be sure to block up the engine under the oil pan flanges or bellhousing bosses to prevent it from tipping over when you install the heads.

Check the new head gaskets from your gasket set and make sure they match the heads. Check the head bolts to be sure you have them all and they're in good condition. If you haven't already done so, clean up the threads of all head bolts with the appropriate die. Clean up the threaded bores in the block for the head bolts with a tap **(see illustration)**. Also clean up the threads in the bores for the intake and exhaust manifolds. Blow all metal debris out of the holes. Clean the head and block gasket surfaces with solvent and blow dry. Make sure all cylinder head locating dowel pins are in place. These dowels are critical - they position the heads properly on the block deck surface. If even one is missing, replace it. Insert the straight end into the block with the chamfered (tapered) end projecting up and tap it into place.

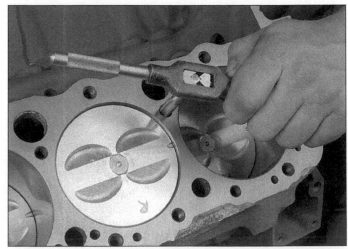

5.42 Clean up the threaded bores in the block for the head bolts - also clean up the threads in the threaded bores in the heads for the intake and exhaust manifolds

5.43 Install the head gaskets with the "Front" designation on the gasket facing up, at the front of the engine - if you install a head gasket backwards or upside down, even if it looks like a good fit, you may block a coolant or lubrication passage, which will cause serious damage to the engine

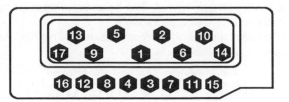

5.44 Head bolt tightening sequence for small-block engines - tighten the bolts smoothly and evenly, in this sequence, to 65 ft-lbs

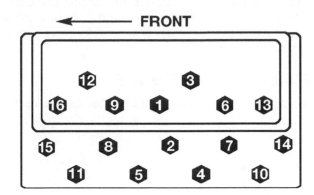

5.45 Head bolt tightening sequence for big-block engines - tighten the bolts smoothly and evenly, in this sequence, to 80 ft-lbs

A few words of advice to owners of pre-1972 engines

If you've got a pre-1972 engine with a high compression ratio, use a hard gasket like Felpro's "Permatorque" or Victor's "Victorcor." These gaskets don't conform to surface irregularities very well, so make sure the block and head gasket surfaces are smooth and distortion-free. **Note:** *Head gaskets are part of a complete engine rebuild gasket set, so make sure you buy a set made by one of the above manufacturers, or some other brand which includes similar hard type head gaskets.*

Because of their small combustion chambers and high compression ratios, pre-1972 engines often experience detonation problems. If you installed low compression pistons, then skip this section. If you didn't, read on.

One way to mitigate detonation, although it may go against manufacturer's recommendations, is to install *two* head gaskets under each head. The extra clearance volume will lower the compression ratio about one point. For instance, if the compression ratio was 11:1, it will be 10:1 with the double gasket setup. When you're talking compression ratio, one point is a lot! Your engine will now be much less likely to detonate.

Installing the head gaskets

It's not necessary to coat most head gaskets. If the gasket manufacturer recommends using no sealant, then don't use any! However, if it's recommended or there are no specific recommendations, use a spray gasket sealant such as High Tack or Copper Coat. Coat both sides of the gaskets evenly and allow it to dry. After it's dry, position the head gaskets on the engine block by locating them on the dowel pins with the FRONT designation on the gaskets at

the front of the engine **(see illustration)**. If you install one of the head gaskets backwards, the coolant passages will be blocked and the head(s) will overheat shortly after startup.

Installing the heads

Note: *If the vehicle is equipped with aluminum heads, see the Haynes Automotive Repair Manual for your particular vehicle for additional procedures and specific torque values.*

Slip a couple of the bottom head bolts into their bolt holes in the head. **Note:** *To prevent coolant leaks, all head bolts must have their threads coated with a non-hardening sealant (such as Permatex no. 2) before installation.* Stretch a rubber band between them to prevent them from protruding out the bottom of the head. With the head gasket in place, lift the cylinder head up and carefully place it on the block. Make sure it engages the locating dowel pins and push it firmly down onto the block. Run in the two bottom head bolts enough to prevent the head from falling off. When you're done with the first head, repeat everything up to this point and install the other head.

Apply engine oil to the threads of the cylinder head bolts and oil them under their heads. Loosely install the cylinder head bolts. Tighten the bolts smoothly and evenly, in sequence, to the specified torque **(see illustrations)**. See the *Haynes Automotive Repair Manual* for your particular vehicle for the torque figures. The head bolts must be tightened in three to five stages as well.

To protect the cylinders from moisture and dirt, install the spark plugs in the heads.

Installing the valve train

Make sure you've got 16 valve lifters, pushrods, rocker arms, fulcrums and nuts or bolts.

Installing the valve lifters

Prime each valve lifter by forcing oil into the side until it squirts out the top oil hole. Coat the outside diameter of each lifter with oil and coat its foot with moly-base grease or engine assembly lube. See Chapter 6 for more information on installing the lifters.

Installing the pushrods

If you're installing new pushrods, they can be installed in any order. If you're installing the old pushrods, install them in the same order in which they were removed, but swap ends, i.e. the end that operated the rocker arm should now be down at the lifter. You can tell which end is which by looking at the wear pattern on each end. The wear pattern on the lifter end is much smaller than the wear pattern at the rocker arm end. Slide each pushrod into a lifter bore and center its lower end in the lifter.

Installing the rocker arms

Apply a coat of oil to the pivot and pushrod-contact points, and apply moly grease to the valve-stem tips. Set each rocker arm on its pushrod and valve tip. If you're installing new rockers, it doesn't matter where they go. But if you're installing the old rockers and pivots, you must keep them together in their original pairs and install them at the same location from which they were removed. Install the pivot balls for the rockers, hand tighten each nut and proceed to Chapter 7, which will show you how to index the crankshaft, adjust the valves and install the rocker arm covers.

6 Overhauling the engine block

Disassembly

Timing chain and sprockets

1   Remove the camshaft sprocket mounting bolts.
2   Pry off the camshaft sprocket and chain with two large screwdrivers or prybars and detach the chain from the crankshaft sprocket **(see illustration)**. Don't lose the pin in the end of the camshaft (if equipped).
3   The crankshaft sprocket can be levered off with two large screwdrivers or prybars.

Lifters

There are several ways to extract the lifters from the bores. Special tools designed to grip and remove lifters are manufactured by many tool companies and are widely available **(see illustration)**, but may not be needed in every case. On newer engines without a lot of varnish buildup, the lifters can often be removed with a small magnet or even with your fingers. A machinist's scribe with a bent end can be used to pull the lifters out by positioning the point under the retainer ring in the top of each lifter. **Caution:** *Don't use pliers to remove the lifters unless you intend to replace them with new ones (along with the camshaft). The pliers may damage the precision machined and hardened lifters, rendering them useless.* On engines with a lot of sludge and varnish, work the lifters up and down, using carburetor cleaner spray to loosen the deposits.

Before removing the lifters, arrange to store them in a clearly labeled box to ensure that they're reinstalled in their original locations. **Note:** *On engines equipped with roller lifters, the guide retainer and guide restrictor must be removed before the lifters are withdrawn.* Remove the lifters

6.1 Using two prybars to remove the camshaft sprocket

6.2 If the lifters are difficult to remove, you may have to remove them using a special lifter puller

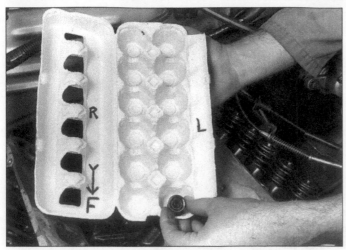

6.3 Be sure to store the lifters in an organized manner to make sure they're reinstalled in their original location

6.4 Removing the oil pump mounting bolt

and store them where they won't get dirty **(see illustration)**.

Oil pump

1   Remove the oil pump mounting bolt **(see illustration)**.
2   Lower the pump and driveshaft from the engine, then separate the driveshaft and nylon sleeve.
3   If the pump will be replaced, separate the pickup tube from the pump.

Piston/connecting rods

1   Completely remove the ridge at the top of each cylinder with a ridge reaming tool **(see illustration)**. Follow the manufacturer's instructions provided with the tool. Failure to remove the ridge before attempting to remove the piston/connecting rod assemblies may result in piston breakage.
2   After the cylinder ridges have been removed, turn the engine upside-down so the crankshaft is facing up.

3   Before the connecting rods are removed, check the endplay with feeler gauges. Slide them between each connecting rod and the crankshaft throw until the play is removed **(see illustration)**. The endplay is equal to the thickness of the feeler gauge(s). If the endplay exceeds the service limit (about 0.023 in. on most models), new connecting rods will be required. If new rods (or a new crankshaft) are installed, the endplay may fall under the specified minimum - about 0.010 in. on most models - (if it does, the rods will have to be machined to restore it - consult an automotive machine shop for advice if necessary). Repeat the procedure for the remaining connecting rods.
4   Check the connecting rods and caps for identification marks **(see illustration)**. If they aren't plainly marked, use a small centerpunch to make the appropriate number of indentations on each rod and cap (1 - 8 on the cylinder they're associated with).
5   Loosen each of the connecting rod cap nuts _-turn at a time until they can be removed by hand. Remove the number one connecting rod cap and bearing insert. Don't drop the bearing insert out of the cap. Slip a short length of plastic or rubber hose over each connecting rod cap bolt to protect the crankshaft journal and cylinder wall as the pis-

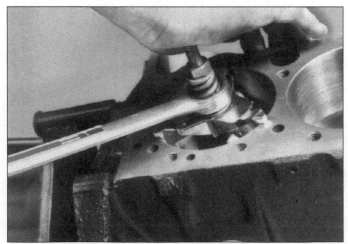

6.5 A ridge reamer is required to remove the ridge from the top of the cylinder - do this before removing the pistons!

6.6 Check the connecting rod side clearance with a feeler guide as shown

6.7 If the connecting rods and caps are not marked to indicate which cylinder they're installed in, mark them with a centerpunch to avoid confusion during reassembly

6.8 To prevent damage to the crankshaft journals and cylinder walls, slip sections of hose over the rod bolts before removing the pistons

ton is removed **(see illustration)**. Push the connecting rod/piston assembly out through the top of the engine. Use a wooden hammer handle to push on the upper bearing insert in the connecting rod. If resistance is felt, double-check to make sure that all of the ridge was removed from the cylinder.

6   Repeat the procedure for the remaining cylinders. After removal, reassemble the connecting rod caps and bearing inserts in their respective connecting rods and install the cap nuts finger tight. Leaving the old bearing inserts in place until reassembly will help prevent the connecting rod bearing surfaces from being accidentally nicked or gouged.

Piston rings

1   Be careful not to nick or gouge the pistons.

2   If a piston ring expander is available, use it to expand and remove the upper two (compression) rings **(see illustration)**.

3   If the expander is not available, use your hands to rotate the upper two off the pistons or break them in half. **Note:** *it's a good idea to wear gloves while removing the*

rings, as the rings may cut you if you're not careful.

4   The three sections of the lower (oil control) ring must be removed by hand.

Crankshaft

1   Before the crankshaft is removed, check the endplay. Mount a dial indicator with the stem in line with the crankshaft and just touching one of the crank throws **(see illustration)**.

2   Push the crankshaft all the way to the rear and zero the dial indicator. Next, pry the crankshaft to the front as far as possible and check the reading on the dial indicator. The distance that it moves is the endplay. If it's greater than about 0.012 in., check the crankshaft thrust surfaces for

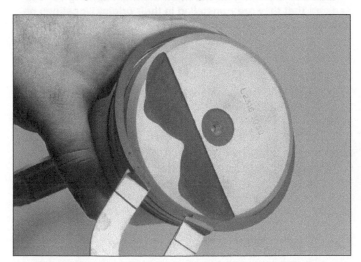

6.9 A ring expander is helpful, but not required, when removing the upper two (compression) rings

6.10 Checking the crankshaft endplay with a dial indicator

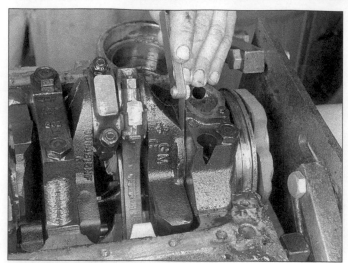

6.11 Checking the crankshaft endplay with a feeler gauge

6.12 Main bearing caps generally have a cast-in-arrow, which points to the front of the engine, and are numbered consecutively from the front of the engine to the rear, but . . .

wear. If no wear is evident, new main bearings should correct the endplay.

3   If a dial indicator isn't available, feeler gauges can be used. Gently pry or push the crankshaft all the way to the front of the engine. Slip feeler gauges between the crankshaft and the front face of the thrust (rear) main bearing to determine the clearance **(see illustration)**.

4   If the engine is equipped with a one-piece rear crankshaft oil seal, unbolt the seal retainer from the rear of the block.

5   Check the main bearing caps to see if they're marked to indicate their locations. They should be numbered consecutively from the front of the engine to the rear. Main bearing caps generally have a cast-in arrow, which points to the front of the engine **(see illustration)**. If they aren't properly marked, mark them with number stamping dies or a centerpunch **(see illustration)**. Loosen each of the main bearing cap bolts - turn at a time each, until they can be removed by hand.

6   Gently tap the caps with a soft-face hammer, then separate them from the engine block. If necessary, use the bolts as levers to remove the caps. Try not to drop the bearing inserts if they come out with the caps.

7   Carefully lift the crankshaft out of the engine. It's a good idea to have an assistant available, since the crankshaft is quite heavy. With the bearing inserts in place in the engine block and main bearing caps, return the caps to their respective locations on the engine block and tighten the bolts finger-tight.

Camshaft

1   Mount a dial indicator so the tip of the indicator is touching the front of the camshaft.

2   Push the camshaft to the rear of the engine and zero the dial.

3   Push the camshaft forward to read the camshaft endplay.

4   The endplay should be approximately in the 0.002 to

6.13 . . . if they are not marked, use a centerpunch or numbered stamping dies to mark the main bearing caps to ensure that they are reinstalled in their original locations on the block (make the punch marks near one of the bolt heads)

0.008 range (for the exact specification, check the *Haynes Automotive Repair Manual* for your particular vehicle). If not within specifications, check the camshaft thrust plate and front of the camshaft for excessive wear once the camshaft is removed.

5   Remove bolts securing the camshaft thrust plate, then separate the plate from the front of the engine block.

6   Carefully pull the camshaft out. Support the cam so the lobes don't nick or gouge the bearings as it's withdrawn **(see illustration)**. If resistance is felt, do not pull harder - check to see where the cam is binding and reposition it.

Core plugs

Remove the core plugs (also known as freeze plugs or soft plugs) from the engine block. To do this, knock one

6.14 Remove the camshaft slowly, supporting it in two places, and make sure the cam lobes and distributor gear don't gouge the camshaft bearings in the block

6.15 Remove the core plugs by tapping one side into the block with a hammer and punch . . .

6.16 . . . and then pull the plug out with pliers

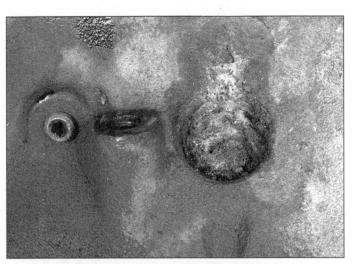

6.17 If the core plugs are rusted badly, side cutters may be used to cut and then pull the plugs out of the block

side of each plug into the block with a hammer and punch **(see illustration)**, then grasp with large pliers and pull out of the block **(see illustration)**.

Oil gallery plugs

Using an Allen head wrench or hex-drive, remove all the oil gallery plugs **(see illustrations)**. The plugs are usually very tight - they may have to be drilled out and the holes retapped. Discard the plugs and use new ones when the engine is reassembled.

Cleaning and inspection

Engine block

1   Using a gasket scraper, remove all traces of gasket material from the engine block. Be very careful not to nick or gouge the gasket sealing surfaces.

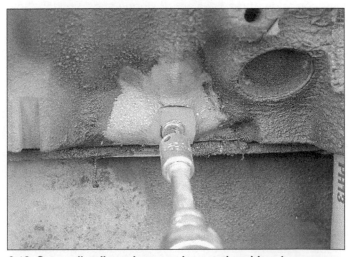

6.18 Some oil gallery plugs can be a real problem to remove – if you get a tough one, try some penetrating oil on the threads

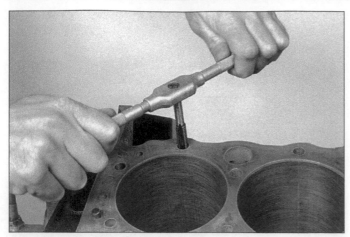

6.19 All bolt holes in the block - particularly the main bearing cap and head bolts holes - should be cleaned and restored with a tap (be sure to remove debris from the holes after this is done)

2   If the engine is extremely dirty it should be taken to an automotive machine shop to be cleaned.

3   After the block is returned, clean all oil holes and oil galleries one more time. Brushes specifically designed for this purpose are available at most auto parts stores. Flush the passages with warm water until the water runs clear, dry the block thoroughly and wipe all machined surfaces with a light, rust preventative oil. If you have access to compressed air, use it to blow out all the oil holes and galleries. **Warning:** *Wear eye protection when using compressed air!*

4   If the block isn't extremely dirty or sludged up, you can do an adequate cleaning job with hot soapy water and a stiff brush. Take plenty of time and do a thorough job. Regardless of the cleaning method used, be sure to clean all oil holes and galleries very thoroughly, dry the block completely and coat all machined surfaces with light oil.

5   The threaded holes in the block must be clean to ensure accurate torque readings during reassembly. Run the proper size tap into each of the holes to remove any rust, corrosion, thread sealant or sludge and to restore any damaged threads **(see illustration)**. If possible, use compressed air to clear the holes of debris produced by this operation. Now is a good time to clean the threads on the head bolts and the main bearing cap bolts as well.

Inspection

1   Double-check to make sure the ridge at the top of each cylinder has been completely removed.

2   Visually check the block for cracks, rust and corrosion. **Note:** *Most cracks are found at the bottom of cylinders, near core plugs, at the main bearing saddles, and between cylinders and water jackets. Check lifter bores for damage. Look for stripped threads in the threaded holes.*

3   It's also a good idea to have the block checked for hidden cracks by an automotive machine shop that has the special equipment to do this type of work. If defects are found, have the block repaired or replaced.

4   Check the cylinder bores for scuffing and scoring.

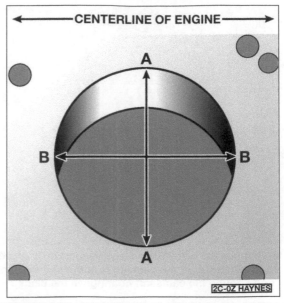

6.20 Measure the diameter of each cylinder at a right angle to the engine centerline (A) and parallel to the engine centerline (B). Out-of-round is the difference between A and B; taper is the difference between A and B at the top of the cylinder and A and B at the bottom of the cylinder

5   Measure the diameter of each cylinder parallel and perpendicular to the crankshaft **(see illustration)** at the top (just under the ridge area), center and bottom of the cylinder bore **(see illustrations)**.

6   The out-of-round specification of the cylinder bore is the difference between the parallel and perpendicular reading. Generally, no cylinder should be more than 0.005 in. out of round (for exact specifications, see the *Haynes Automotive Repair Manual* for your particular vehicle)

7   The taper of the cylinder is the difference between the bore diameter at the top and the diameter at the bottom. Generally, taper should not exceed 0.010 in. (for exact specifications, see the *Haynes Automotive Repair Manual* for your particular vehicle).

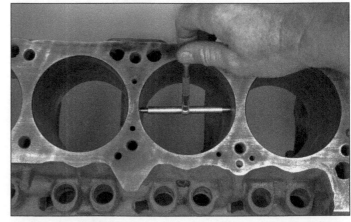

6.21a The ability to "feel" when the telescoping gauge is at the correct point will be developed over time, so work slowly and repeat the check until you're satisfied the bore measurement is accurate

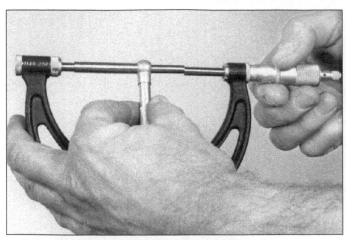

6.21b The gauge is then measured with a micrometer to determine the bore size

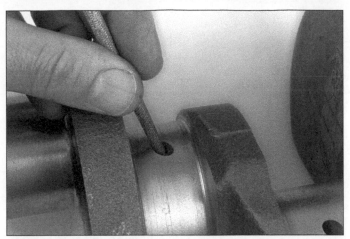

6.22 The oil holes should be chamfered so sharp edges don't gouge or scratch the new bearings

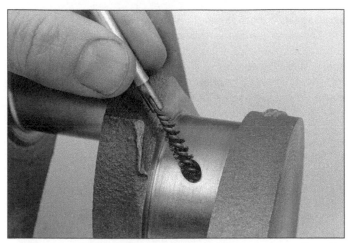

6.23 Use a wire or stiff plastic bristle brush to clean the oil passages in the crankshaft

6.24 Measure at several points around each journal's circumference, then measure at each end - this will help you identify out-of-round and taper

Crankshaft

1   Clean the crankshaft with solvent and dry it with compressed air (if available). Be sure to clean the oil holes with a stiff brush and flush them with solvent **(see illustration)**. Chamfer the oil holes to remove sharp edges **(see illustration)**. Check the main and connecting rod bearing journals for uneven wear, scoring, pits and cracks. Visually check the rest of the crankshaft for cracks and other damage.

2   Using a micrometer, measure the diameter of the main and connecting rod journals **(see illustration)**. Check the *Haynes Automotive Repair Manual* for your particular vehicle for the correct specifications. By measuring the diameter at a number of points around each journal circumference, you'll be able to determine whether or not the journal is out-of-round. Take the measurement at each end of the journal, near the crank throws, to determine if the journal is tapered. Generally, taper and out-of-round should not exceed about 0.0005 inch.

3   Check the crankshaft runout by placing the crankshaft in the block supported by the end bearings only. Using a micrometer placed against the center main bearing journal,

6.25 Using a dial indicator to measure the crankshaft runout

rotate the crankshaft to read the runout **(see illustration)**. Generally, runout should not exceed about 0.004 in.

4   If the crankshaft journals are damaged, tapered, out-of-

6.26 Rubbing a penny lengthwise on each journal will reveal its condition - if copper rubs off and is embedded in the crankshaft, the journals should be reground

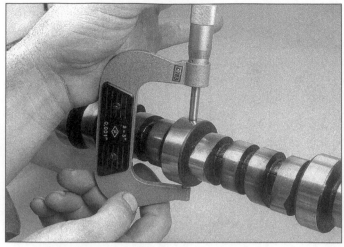

6.27 Measuring the camshaft bearing journals

round or worn excessively, have the crankshaft reground by an automotive machine shop. Be sure to use the correct oversize bearing inserts if the crankshaft is reconditioned.

5   Check the oil seal contact surfaces on the crankshaft. If they're scratched, nicked or otherwise damaged, the oil seals may leak when the engine is reassembled. Repair may be possible (ask at an automotive machine shop).

6   If the dimensions of the crankshaft are OK, rub a penny across each journal several times **(see illustration)**. If a journal picks up copper from the penny, it's too rough and must be reground.

Camshaft

1   After the camshaft has been removed from the engine, cleaned with solvent and dried, inspect the bearing journals for uneven wear, pits and galling. If the journals are damaged, the bearing inserts in the block are probably damaged as well. Replace both the camshaft and bearings if this is the case. Normally, if the journals are not damaged

and have a smooth finish, they are OK.

2   Measure the inside diameter of each camshaft bearing and record the results (take two measurements, 90 degrees apart, at each bearing).

3   Measure the camshaft bearing journals with a micrometer **(see illustration)**. Compare the measurements with the specifications in the *Haynes Automotive Repair Manual* for your particular vehicle. If they're less than specified, the camshaft should be replaced with a new one.

4   Subtract the bearing journal diameters from the corresponding bearing inside diameter measurements to obtain the oil clearance. Generally, the oil clearance should not exceed about 0.006 in.

5   Check the camshaft lobes for heat discoloration, score marks, chipped areas and pitting. If any of these conditions are present on any lobe, replace the camshaft.

6   If you have lobe lift specifications available for the camshaft, measure it as follows.

7   Using a micrometer, measure the lobe at its greatest dimension and 90-degrees from the greatest dimension **(see illustrations)**.

6.28a Measuring the camshaft lobe at its greatest dimension

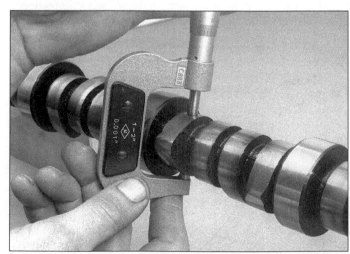

6.28b Measuring the camshaft lobe at its smallest dimension

6.29 If the lifter is flat or pitted it must be replaced

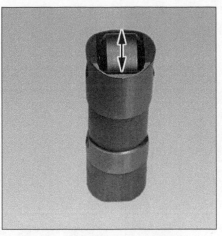

6.30 The roller on roller lifters must turn freely - check for wear and excessive play as well

6.31a The piston ring grooves can be cleaned with a special tool, as shown here . . .

8   Camshaft lobe lift is the difference between these two measurements. If it's less than specified for the camshaft, replace the camshaft.

Lifters

Conventional lifters

1   Clean the lifters with solvent and dry them thoroughly without mixing them up.
2   Check each lifter wall, pushrod seat and foot for scuffing, score marks and uneven wear. Each lifter foot (the surface that rides on the cam lobe) must be slightly convex, although this can be difficult to determine by eye.
3   If the base of the lifter is flat or concave **(see illustration)**, the lifters and camshaft must be replaced.
4   If the lifter walls are damaged or worn (which isn't very likely), inspect the lifter bores in the engine block as well.
5   If new lifters are being installed, a new camshaft must also be installed.
6   If a new camshaft is installed, then use new lifters as well. Never install used lifters unless the original camshaft is used and the lifters can be installed in their original locations!

Roller lifters

7   Check the rollers carefully for wear and damage and make sure they turn freely without excessive play **(see illustration)**.
8   The inspection procedure for the sides (walls) of conventional lifters also applies to roller lifters.
9   Unlike conventional lifters, used roller lifters can be reinstalled with a new camshaft and the original camshaft can be used if new lifters are installed.

Pistons/connecting rods

1   Before the inspection process can be carried out, the piston/rods must be cleaned and the original piston rings removed from the pistons. **Note:** *Always use new piston*

6.31b . . . or section of a broken ring

rings when the engine is reassembled.
2   Scrape all traces of carbon from the top of the piston. A hand-held wire brush or a piece of fine emery cloth can be used once the majority of the deposits have been scraped away. Do not, under any circumstances, use a wire brush mounted in a drill motor to remove deposits from the pistons. The piston material is soft and may be eroded away by the wire brush.
3   Use a piston ring groove cleaning tool to remove carbon deposits from the ring grooves. If a tool isn't available, a piece broken off the old ring will do the job. Be very careful to remove only the carbon deposits - don't remove any metal and don't nick or scratch the sides of the ring grooves **(see illustrations)**.
4   Once the deposits have been removed, clean the piston/rod assemblies with solvent and dry them with compressed air (if available). Make sure the oil return holes in the back sides of the ring grooves are clear.
5   If the pistons and cylinder walls aren't damaged or worn excessively, and if the engine block is not rebored, new pistons won't be necessary. Normal piston wear

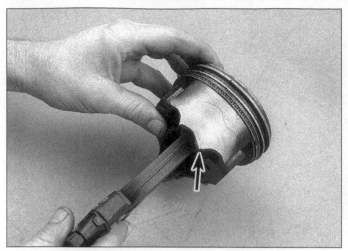

6.32 The piston has a broken skirt (arrow)

6.33 This is what happens when a valve drops into the combustion chamber

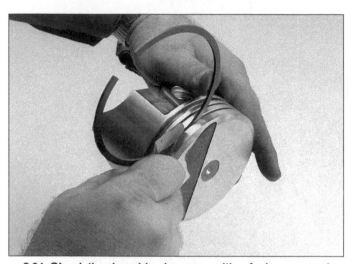

6.34 Check the ring side clearance with a feeler gauge at several points around the groove

6.35 Measure the piston diameter at a 90-degree angle to the piston pin and in line with it

appears as even, vertical wear on the piston thrust surfaces and slight looseness of the top ring in its groove. New piston rings should always be used when an engine is rebuilt.

6   Carefully inspect each piston for cracks around the skirt **(see illustration)**, at the pin bosses and at the ring lands.

7   Look for scoring and scuffing on the thrust faces of the skirt, holes in the piston crown **(see illustration)** and burned areas at the edge of the crown. If the skirt is scored or scuffed, the engine may have been suffering from over-heating and/or abnormal combustion. The cooling and lubrication systems should be checked thoroughly. A hole in the piston crown is an indication that abnormal combus-tion (preignition or detonation) was occurring, the piston hit an open valve (improper valve timing) or a foreign object was in the combustion chamber. If any of the above prob-lems exist, the causes must be corrected or the damage will occur again.

8   Corrosion of the piston, in the form of small pits, indi-cates that coolant is leaking into the combustion chamber and/or the crankcase. Again, the cause must be corrected or the problem may persist in the rebuilt engine.

9   Measure the piston ring side clearance by laying a new piston ring in each ring groove and slipping a feeler gauge in beside it **(see illustration)**. Check the clearance at three or four locations around each groove. Be sure to use the correct ring for each groove; they are different. If the side clearance is greater than about 0.004 in., new pistons will have to be used.

10   Check the piston-to-bore clearance by measuring the bore and the piston diameter. Make sure the pistons and bores are correctly matched. Measure the piston across the skirt, at a 90-degree angle to and in line with the piston pin **(see illustration)**. Subtract the piston diameter from the bore diameter to obtain the clearance. Generally, it should not be more than about 0.003 in. (for exact specifications, check the *Haynes Automotive Repair Manual* for your par-ticular vehicle). If it's greater than specified, the block will have to be rebored and new pistons installed.

11   Check the piston-to-rod clearance by twisting the pis-

Overhauling the engine block

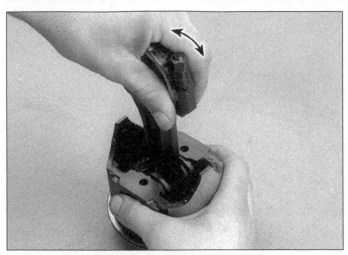

6.36 Check for play between the piston and the rod assembly by twisting the rod against the pin

ton and rod in opposite directions **(see illustration)**. Any noticeable play indicates excessive wear, which must be corrected. The piston/connecting rod assemblies should be taken to an automotive machine shop to have the pistons and rods rebored and new pins installed.

12 If the pistons must be removed from the connecting rods for any reason, they should be taken to an automotive machine shop. While they are there, have the connecting rods checked for bend and twist, since machine shops have special equipment for this purpose. **Note:** *Unless new pistons and/or connecting rods must be installed, do not disassemble the pistons and connecting rods.*

13 Check the connecting rods for cracks and other damage. Temporarily remove the rod caps, lift out the old bearing inserts, wipe the rod and cap bearing surfaces clean and inspect them for nicks, gouges and scratches. After checking the rods, replace the old bearings, slip the caps into place and tighten the nuts finger tight.

Bearings

1 The main and connecting rod bearings should be replaced with new ones during the engine overhaul, but the old bearings should be retained for close examination, as they may reveal valuable information about the condition of the engine **(see the bearing condition chart on pages 6-13 through 6-16)**.

2 When examining the crankshaft bearings, remove them from the engine block, the main bearing caps, the connecting rods and the rod caps and lay them out on a clean surface in the same general position as their location in the engine. This will enable you to match any bearing problems with the corresponding crankshaft journal.

3 Bearing failure occurs because of lack of lubrication, the presence of dirt or other foreign particles, overloading the engine and corrosion. Regardless of the cause of bearing failure, it must be corrected before the engine is reassembled to prevent it from happening again.

4 Dirt and other foreign particles get into the engine in a

variety of ways. It may be left in the engine during assembly, or it may pass through filters or the PCV system. It may get into the oil, and from there into the bearings. Metal chips from machining operations and normal engine wear are often present. Abrasives are sometimes left in engine components after reconditioning, especially when parts are not thoroughly cleaned using the proper cleaning methods. Whatever the source, these foreign objects often end up embedded in the bearing material and are easily recognized. Large particles will not embed in the bearing and will score or gouge the bearing and journal. The best prevention for this cause of bearing failure is to clean all parts thoroughly and keep everything spotlessly clean during engine assembly. Frequent and regular engine oil and filter changes are also recommended.

5 Lack of lubrication (or lubrication breakdown) has a number of interrelated causes. Excessive heat (which thins the oil), overloading (which squeezes the oil from the bearing face) and oil leakage or throw-off (from excessive bearing clearances, worn oil pump or high engine speeds) all contribute to lubrication breakdown. Blocked oil passages, which usually are the result of misaligned oil holes in a bearing, will also oil starve a bearing and destroy it. When lack of lubrication is the cause of bearing failure, the bearing material is wiped or extruded from the steel backing of the bearing. Temperatures may increase to the point where the steel backing turns blue from overheating.

6 Driving habits can have a definite effect on bearing life. Full throttle, low speed operation (lugging the engine) puts very high loads on bearings, which tends to squeeze out the oil film. These loads cause the bearings to flex, which produces fine cracks in the bearing face (fatigue failure). Eventually the bearing material will loosen in pieces and tear away from the steel backing. Short-trip driving leads to corrosion of bearings because insufficient engine heat is produced to drive off the condensed water and corrosive gases. These products collect in the engine oil, forming acid and sludge. As the oil is carried to the engine bearings, the acid attacks and corrodes the bearing material.

7 Incorrect bearing installation during engine assembly will lead to bearing failure as well. Tight-fitting bearings leave insufficient bearing oil clearance and will result in oil starvation. Dirt or foreign particles trapped behind a bearing insert result in high spots on the bearing which lead to failure.

Oil pump

1 If the engine is worn and metal chips have been floating in the oil, chances are the oil pump is also worn and damaged. Unless the oil pump has been recently replaced, is known to be in good condition and passes the following checks, don't even bother checking it - replace it with a new one. The pump is relatively inexpensive when compared with the expense of the freshly overhauled engine.

2 With the oil pump pickup in a container of clean solvent, turn the driveshaft. After two or three revolutions, the solvent should flow out of the pump's pressure port (the

6-11

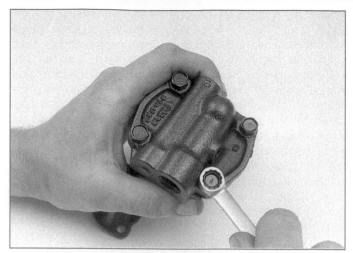

6.38 Remove the bolts retaining the oil pump cover

6.39 Inspect the pump cover for scoring and pitting - this cover shows wear; the pump should be replaced

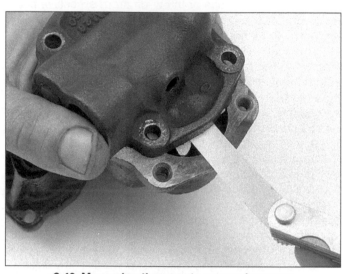

6.40 Measuring the gear-to-cover clearance

hole in the center of the pump mounting flange). Replace the pump if it does not perform as described.

3   Inspect the driveshaft for damage.

4   Remove the cover from the end of the pump **(see illustration)**.

5   Inspect the inside of the cover for groves and wear **(see illustration)**. Replace the pump if either is present.

6   If the cover is okay, position it over the pump and measure the clearance between the gear assembly and cover **(see illustration)**. This will give you the gear assembly end clearance. Generally, it should not exceed about 0.003 in.

7   Remove the gears.

8   Inspect the gear and gear shafts for wear.

9   Inspect the gear housing **(see illustration)**.

10   Use a screwdriver and push in on the relief valve to be sure it is free and has spring pressure **(see illustration)**.

11   If any of the above checking procedures indicates a problem, the pump should be replaced.

6.41 Inspect the gear housing for scoring, pitting and wear - this housing shows normal wear

6.42 Push on the pressure relief valve to make sure it's free and has spring pressure

Service record

Date	Mileage	Work performed

ENGINE BEARING ANALYSIS

Debris

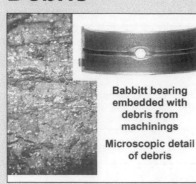

Babbitt bearing embedded with debris from machinings

Microscopic detail of debris

Microscopic detail of gouges

Overplated copper alloy bearing gouged by cast iron debris

Aluminum bearing embedded with glass beads

Microscopic detail of glass beads

Damaged lining caused by dirt left on the bearing back

Misassembly

Result of a lower half assembled as an upper - blocking the oil flow

Excessive oil clearance is indicated by a short contact arc

Polished and oil-stained backs are a result of a poor fit in the housing bore

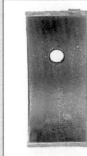

Result of a wrong, reversed, or shifted cap

Overloading

Damage from excessive idling which resulted in an oil film unable to support the load imposed

Damaged upper connecting rod bearings caused by engine lugging; the lower main bearings (not shown) were similarly affected

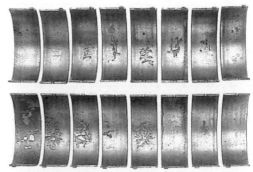

The damage shown in these upper and lower connecting rod bearings was caused by engine operation at a higher-than-rated speed under load

Misalignment

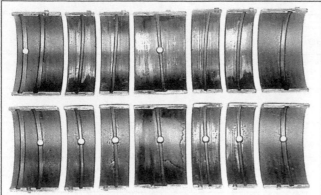

A warped crankshaft caused this pattern of severe wear in the center, diminishing toward the ends

A poorly finished crankshaft caused the equally spaced scoring shown

A tapered housing bore caused the damage along one edge of this pair

A bent connecting rod led to the damage in the "V" pattern

Lubrication

Result of dry start: The bearings on the left, farthest from the oil pump, show more damage

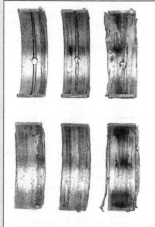

Result of a low oil supply or oil starvation

Severe wear as a result of inadequate oil clearance

Corrosion

Microscopic detail of corrosion

Corrosion is an acid attack on the bearing lining generally caused by inadequate maintenance, extremely hot or cold operation, or inferior oils or fuels

Microscopic detail of cavitation

Example of cavitation - a surface erosion caused by pressure changes in the oil film

Damage from excessive thrust or insufficient axial clearance

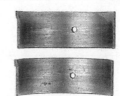

Bearing affected by oil dilution caused by excessive blow-by or a rich mixture

Service record

Date	Mileage	Work performed

6.43 Uneven piston wear like this indicated a bent connecting rod

6.44 A machinist checks for a bent connecting rod

6.45 A bore gauge being used to check the main bearing bore

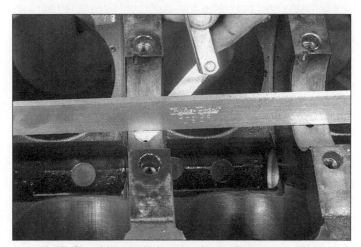

6.46 Checking main bearing bore alignment with a straightedge and feeler gauge

Machine shop procedures

Inspections

Connecting rod bend/twist

The following procedures are normally carried out by an automotive machine shop, since special equipment and skills are required. The information included here will help you communicate with the shop that performs the work.

Inspect the pistons for uneven wear marks. If the pistons have uneven wear marks, it's very possible the rod is bent or twisted (see illustration). Take the piston/connecting rod assemblies to an automotive machine shop so they can be inspected on a special jig that checks for connecting rod bend and twist (see illustration).

Main bearing bore and alignment

Bore

The main bearing bore can be checked with the main bearing caps installed and tightened to the correct specification. General specifications are as follows:

Small-block with two-bolt main bearing
    caps - 75 ft-lbs.
Small-block with four-bolt caps - 65 ft lbs.
Big-block with two-bolt caps - 95 ft-lbs.
Big block with four-bolt caps - 110 ft lbs.

The bore size and out-of-round are determined by using an inside micrometer or bore gauge and micrometer (see illustration). If the diameter is not within specifications (listed in your vehicle's Haynes manual or available from an automotive machine shop), the block must be align-bored by a machine shop.

Alignment

An automotive machine shop can check the bore alignment. Alignment is checked by placing a precision straightedge along the centerline of at least three main bearing

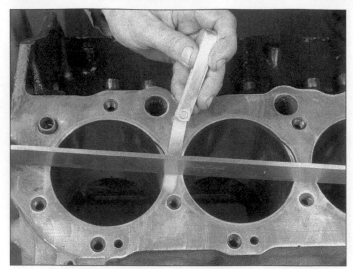

6.47 A precision straightedge and feeler gauge are used to check the deck for warpage

saddles. A feeler gauge is then used to check for clearance between the straightedge and main bearing saddles **(see illustration)**. The bock must be align bored if clearance is more than half of the recommended bearing clearance.

The alignment can also be checked using a test arbor 0.001 inch less than the bearing bore inserted and bolted in place of the crankshaft. The alignment is OK if the test arbor can be rotated using a 12-inch handle.

Deck warpage

The block deck can also be checked for warpage by an automotive machine shop. It's checked by placing a precision straightedge across the gasket surface of the deck. A feeler gauge is used to measure the clearance between the straightedge and the deck. Measurements are taken across the width and length of cylinder block at several points **(see illustration)**.

If total deck warpage exceeds 0.006 in. overall (or 0.003 in. in any six inches), the block deck must be machined flat. If 0.020 in. or more is machined from the cylinder head surface and block deck combined, the intake manifold will also have to be machined to accommodate for a smaller valley between the heads (see Chapter 5).

Checking for cracks

Note: *These procedures must be performed by an automotive machine shop, since special equipment and skills are required. A brief overview is included here to help you communicate with the machine shop that performs the work.*

Magnaflux

This type of inspection is restricted to ferrous (magnetic) materials, most commonly cylinder blocks and crankshafts. It is done by magnetizing the component or a localized area of the component. Metal dust is applied to the magnetized area. The dust is sucked down into cracks by magnetism, causing the cracks to become visible.

6.48 An engine block being bored

Pressure

Components with cavities can be tested for cracks using a pressure tester. This type of testing is done by plugging all but one of the holes of the component and injecting air or water into the open passage.

When using water, cracks are indicated by the appearance of wet or damp areas. When air is used, it is necessary to spray the surface with soapy water and look for bubbles.

Zyglo

This procedure is commonly used on non-ferrous materials (like aluminum). The component is coated with a special fluorescent dye and often warmed to expand any cracks that might be present. The dye seeps into the expanded cracks. The surface is then thoroughly cleaned, and a developing solution is applied to the surface. Dye seeping out of a crack is highlighted by the developing solution and is easily visible in a darkened room when inspected with a black light.

Sound waves

This is an expensive process. Because of cost, it's usually restricted to expensive racing engine parts. It uses ultrasonic waves to find internal cracks in components that cannot be identified in any other way.

X-ray

This is another expensive inspection that is used to detect both internal and external cracks and flaws. It works much like medical X-rays and requires expensive equipment and careful analysis. Again, it's normally used for racing engines.

Repairs

Cylinder reboring

Cylinder reboring is done with a piece of equipment especially designed for this process, usually called a boring bar **(see illustration)**. It consists of a power-driven cutting tool in an arbor. The arbor is aligned on the cylinder center near the bottom of the cylinder bore where the least wear has occurred.

When centered, the boring machine is clamped in position, often on the block deck surface (if their is any deck warpage it must be milled flat). The arbor is raised and fitted with a cutting tool. With the main bearing caps tightened to specification, the cutting tool is adjusted to the diameter desired and the cylinder is then bored from the top down. Several passes, rough and fine, may be required to enlarge the cylinder to the required size. The cylinder is bored to .0005 to .002 in. less than required. Cylinder size and required wall finish are completed by honing.

Camshaft bearing replacement

After removing the camshaft rear bearing bore plug, a special tool is used to remove and install the camshaft bearings. It is essential the front bearing be installed to the correct depth and oil holes in all the bearings line up with the oil holes in the block.

Crankshaft grinding

Lightly damaged journals can be corrected by grinding the journals on true centers **(see illustration)**. If the shaft

6.49 A crankshaft having a main bearing journal ground

has bends, the machine shop will straighten them before grinding.

If the standard production crankshaft is damaged beyond grinding limits, it should be replaced. More expensive racing crankshafts, or crankshafts that are to be modified, can have the journals built up by welding or by special metal spray techniques. They are then reground. This process is expensive and is only done when it is less costly than purchasing a new crankshaft.

Camshaft grinding

Camshafts can be reground on cam grinders. This can be done as a repair or a means of modifying the cam lobe (the lobe can be built up by welding before grinding). Since it is generally more expensive to grind a camshaft than to purchase a new camshaft, grinding is only done on special camshafts and on some low-production camshafts.

Repairing cracks in the engine block

Cracks in the block will allow leaks or will not support engine loads. The location of the crack and cost of repair are factors to consider when deciding whether to repair the crack or replace the block. Cracks on the exterior of a coolant passage are usually easily repairable by welding. Cracks in the webs between the main bearings are not repairable.

Choosing piston rings and honing the cylinders

Choosing rings

Generally, three types of rings are available. The type you choose will determine how coarsely you hone your cylinders and, also, how long the break-in period will be after the engine is overhauled.

a) *Cast iron - This ring breaks in fast, but doesn't last as long as other types of rings.*

b) *Moly faced - This type of ring is designed to retain a maximum amount of oil between the ring and the cylinder wall. It costs more and takes longer to break in than a cast iron ring, but it lasts longer.*

c) *Chrome faced - This ring is designed to be used in industrial applications where a lot of dirt will be inhaled by the engine. The break-in period for these rings is usually longer than other types. They are also more expensive than cast-iron rings.*

Honing the cylinders

1   If the cylinders are bored, the cylinder should be bored to 0.0005 to 0.002 in. less than required. The cylinder size and required wall finish should be completed by honing **(see illustration on next page)**.

2   If the cylinders are not bored, you can do the honing yourself or have an automotive machine shop do it for you

6.50 If the cylinders are bored, the machine shop will normally do the honing for you, on a machine like this

6.51 You can also hone the cylinders using a drill motor and a special cylinder hone

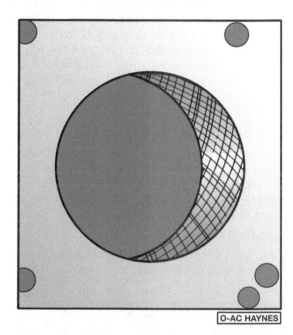

6.52 The cylinder hone should leave a smooth, crosshatch pattern with the lines intersecting at approximately a 60-degree angle

(usually at a reasonable charge) **(see illustration)**. The type of ring being used will determine the grit of the honing stone required for the finish. If cast-iron or chrome-faced rings are being used, the honing grooves for oil retention should be deeper. For the finish honing, a 280-grit stone should be used.

3 If moly rings are being used, the honing grooves for oil retention should be shallow. For the finish honing, a 400-grit stone should be used.

4 To hone the cylinders, we recommend the following:

a) *Install the main bearing caps and tighten the bolts to the specified torque.*

b) *Mount the hone in a drill motor, compress the stones (if necessary) and slip it into the first cylinder. Be sure to wear safety goggles or a face shield!*

c) *Lubricate the cylinder with equal parts of 20 Wt. oil and kerosene, turn on the drill and move the hone up-and-down in the cylinder at a pace that will produce a crosshatch pattern on the cylinder walls. Ideally, the crosshatch lines should intersect at approximately a 60-degree angle* **(see illustration)**. *Be sure to use plenty of lubricant and don't take off any more material than is necessary.*

d) *Don't withdraw the hone from the cylinder while it's running. Instead, shut off the drill and continue moving the hone up and down in the cylinder until it comes to a complete stop, then compress the stones and withdraw the hone.*

e) *Wipe the oil out of the cylinder and repeat the procedure for the remaining cylinders.*

f) *After the honing job is complete, chamfer the top edges of the cylinder bores with a small file so the rings won't catch when the pistons are installed. Be very careful not to nick the cylinder walls with the end of the file!*

6.53 A large socket on an extension can be used to drive the new core plugs into the bores

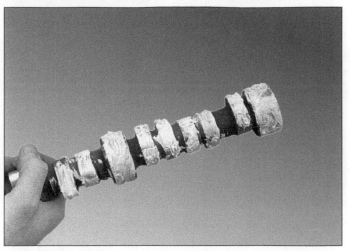

6.54 Apply moly-base grease or engine assembly lube to the camshaft lobes and journals prior to installation

g) *The entire engine block must be washed again very thoroughly with warm, soapy water to remove all traces of the abrasive grit produced during the honing operation.* **Note:** *The bores can be considered clean when a white cloth - dampened with clean engine oil - used to wipe them down doesn't pick up any more honing residue, which will show up as gray areas on the cloth.*

h) *After rinsing, dry the block and apply a coat of light rust preventive oil to all machined surfaces. Wrap the block in a plastic trash bag to keep it clean and set it aside until reassembly.*

Reassembly

Core plugs and oil gallery plugs

1   Coat the plug and bore lightly with an oil-resistant sealant.

2   Cup-type core plugs are installed with the flanged edge out.

3   The plug must be driven into the bore using a tool that does not contact the flange **(see illustration)**.

4   The flanged edge of the plug must rest just below the chamfered edge of the bore in the block.

5   Oil gallery plugs are installed in the same manner, except, after installing the plugs, it's a good idea to stake the edge of the cylinder block around the plug with a hammer and chisel. This will assure the plugs stay in place. Threaded plugs can be screwed into place after coating the threads with sealant.

Camshaft

1   Apply engine assembly lube to the lobes, bearing journals of the camshaft and the camshaft bearings **(see illustration)**.

2   While being very careful not to damage the bearings, insert the camshaft into the block. Support the camshaft in

two places as you insert it.

3   Install the thrust plate and tighten the retaining bolts. Check the endplay (see the procedure earlier in this Chapter).

Crankshaft/main bearings/rear main oil seal

1   Crankshaft installation is the first major step in engine reassembly. It's assumed at this point that the engine block and crankshaft have been cleaned, inspected and reconditioned.

2   Position the engine with the bottom facing up.

3   Remove the main bearing cap bolts and lift out the caps. Lay them out in the proper order to ensure correct installation.

4   If they're still in place, remove the old bearing inserts from the block and the main bearing caps. Wipe the main bearing surfaces of the block and caps with a clean, lint-free cloth. They must be kept spotlessly clean.

5   Clean the back sides of the new main bearing inserts and lay one bearing half in each main bearing saddle of the block. Lay the other bearing half from each bearing set in the corresponding main bearing cap. Make sure the tab on the bearing insert fits into the recess in the block or cap. Also, the oil holes in the block must line up with the oil holes in the bearing insert. **Caution:** *Do not hammer the bearings into place and don't nick or gouge the bearing faces.* 6

The flanged thrust bearing must be installed in the rear (fifth) cap and saddle.

7   Clean the faces of the bearings in the block and the crankshaft main bearing journals with a clean, lint-free cloth.

8   Carefully lay the crankshaft in position in the main bearings.

9   Before the crankshaft can be permanently installed, the main bearing oil clearance must be checked.

10   Trim pieces of the appropriate size of Plastigage (they must be slightly shorter than the width of the main bearings)

6.55 Lay the Plastigage strips (arrow) on the main bearing journals, parallel to the crankshaft centerline

6.56 Compare the width of the crushed Plastigage to the scale on the envelope to determine the main bearing oil seal clearance (always take the measurement at the widest point of the Plastigage); be sure to use the correct scale - standard and metric ones are included

and place one piece on each crankshaft main bearing journal, parallel with the journal axis **(see illustration)**.

11 Clean the faces of the bearings in the caps and install the caps in their respective positions (don't mix them up) with the arrows pointing toward the front of the engine. Don't disturb the Plastigage.

12 Starting with the center main and working out toward the ends, tighten the main bearing cap bolts, in three steps, to the specified torque. General specifications are listed in the main bearing bore check procedure earlier in this Chapter. **Note:** *Don't rotate the crankshaft at any time during this operation.*

13 Remove the bolts and carefully lift off the main bearing caps. Keep them in order. Don't disturb the Plastigage or rotate the crankshaft. If any of the main bearing caps are difficult to remove, tap them gently from side-to-side with a soft-face hammer to loosen them.

14 To obtain the main bearing oil clearance, compare the width of the crushed Plastigage on each journal to the scale printed on the Plastigage container **(see illustration)**. Generally, the clearance should be about 0.001 to 0.002 in. (for exact specifications, see the *Haynes Automotive Repair Manual* for your particular vehicle).

15 If the clearance is not as specified, the bearing inserts may be the wrong size (which means different ones will be required). Before deciding that different inserts are needed, make sure that no dirt or oil was between the bearing inserts and the caps or block when the clearance was measured. If the Plastigage is noticeably wider at one end than the other, the journal may be tapered.

16 If the clearances are OK, carefully scrape all traces of the Plastigage material off the main bearing journals and/or the bearing faces. Don't nick or scratch the bearing faces.

17 Carefully lift the crankshaft out of the engine.

18 If you are working on an engine equipped with a one-piece rear main seal, go to Step 28. If you are working on an engine with a two-piece seal (rope-type or lip-type), proceed as follows:

Rope-type seal

19 Lay one seal section on edge in the seal groove in the block and push it into place with your thumbs. Both ends of the seal should extend out of the block slightly **(see illustration)**.

20 Seat it in the groove by rolling a large socket or piece of bar stock along the entire length of the seal **(see illustration)**. As an alternative, push the seal very carefully into place with a wooden hammer handle.

21 Once you are satisfied that the seal is completely seated in the groove, trim off the excess on the ends with a single-edge razor blade or razor knife (the seal ends must be flush with the block-to-cap mating surfaces **(see illustrations)**. Make sure that no seal fibers get caught between the block and cap.

22 Repeat the entire procedure to install the other half of the seal in the bearing cap. Apply a thin film of engine

6.57 When correctly installed, the ends of the rope-type seal should extend out of the block

6.58 Seat the seal in the groove, but do not depress it below the bearing surface (the seal must contact the crankshaft journal)

6.59 Trim the ends flush with the block . . .

assembly lube to the edge of the seal (where it contacts the crankshaft) **(see illustration)**.

23 During the final installation of the crankshaft (after the main bearing oil clearances have been checked with Plasti-gage), apply a thin, even film of anaerobic-type gasket sealant to the areas of the rear main bearing cap indicated in the accompanying illustrations. **Caution:** *Do not get any sealant on the bearing or seal faces.*

Neoprene lip-type seal (two piece)

24 Inspect the bearing cap and engine block mating surfaces and seal grooves for nicks, burrs and scratches. Remove any defects with a fine file or deburring tool.

25 Install one seal section in the block with the lip facing the *front* of the engine **(see illustrations)**. Leave one end protruding from the block approximately _ to 3/8-inch and make sure it is completely seated.

6.60 . . . but leave the inner edge (arrow) protruding slightly

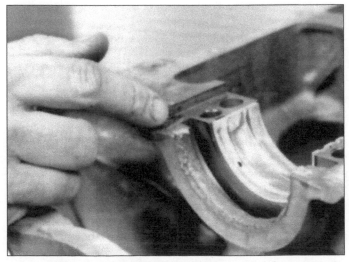

6.61 Lubricate the seal with assembly lube or moly-based grease

6.62 When applying the sealant, be sure it gets into the corner and onto the vertical cap-to-block mating surface or oil leaks will result

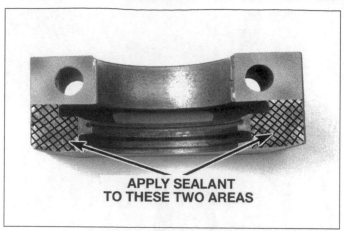

6.63 Apply anaerobic-type gasket sealant to the shaded areas of the rear main bearing cap

6.64 Using the tool that comes with the seal like a "shoehorn", slide the seal sections onto the bearing cap and block . . .

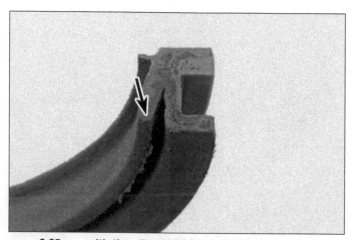

6.65 . . . with the oil seal lip (arrow) facing the front of the engine

6.66 Apply engine assembly lube to the bearing surfaces of the crankshaft

26 Repeat the procedure to install the remaining seal half in the rear main bearing cap. In this case, leave the opposite end of the seal protruding from the cap approximately the same distance the block seal is protruding from the block.

27 During the final installation of the crankshaft (after the main bearing oil clearances have been checked with Plastigage) apply a thin, even film of anaerobic-type gasket sealant to the areas of the cap or block indicated in **illustration 6.63**. **Caution:** *Do not get any sealant on the bearing face, crankshaft journal or seal lip.* Also, lubricate the seal lips with moly-based grease or engine assembly lube.

28 Clean the bearing faces in the block, then apply a thin, uniform layer of clean moly-base grease or engine assembly lube to each of the bearing surfaces **(see illustration)**. Be sure to coat the thrust faces as well as the journal face of the rear (thrust) bearing.

29 Make sure the crankshaft journals are clean, then lay the crankshaft back in place in the block. Clean the faces of the bearings in the caps, then apply lubricant to them. Install the caps in their respective positions with the arrows pointing toward the front of the engine. Install the bolts.

30 Tighten all except the rear cap bolts (the one with the thrust bearing) to the specified torque (work from the center out and approach the final torque in three steps).

31 Tighten the third cap bolts finger tight.

32 Pry the crankshaft forward against the thrust surface of the upper half of the bearing.

33 Hold the crankshaft forward and pry the thrust bearing cap to the rear. While retaining the forward pressure on the crankshaft, tighten the cap bolts to specification.

34 Retighten all main bearing cap bolts to the specified torque, starting with the center main and working out toward the ends.

35 Rotate the crankshaft a number of times by hand to check for any obvious binding.

36 Check the crankshaft endplay with a feeler gauge or a dial indicator (see the procedure earlier in this Chapter). The endplay should be correct if the crankshaft thrust faces aren't worn or damaged and new bearings have been installed.

37 On small-block engines with a one-piece seal **(see illustration)**, clean the crankshaft and seal retainer bore with lacquer thinner or acetone. Check the seal contact sur-

6.67 To remove the seal from the housing, insert the tip of the screwdriver into each notch and pry out the seal

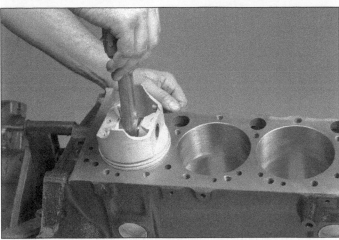

6.68 When checking piston ring end gap, the ring must be square in the cylinder bore (this is done by pushing the ring down with the top of the piston as shown)

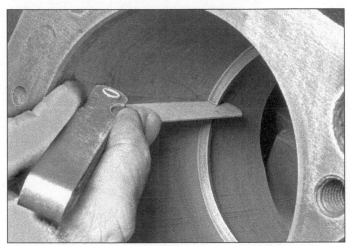

6.69 With the ring square in the cylinder, measure the end gap with a feeler gauge

6.70 If the end gap is too small, clamp a file in a vise and file the ring ends (from the outside in only) to enlarge the gap slightly

face very carefully for scratches and nicks that could damage the new seal lip and cause oil leaks.

38 Make sure the bore is clean, then apply a thin coat of engine oil to the outer edge of the new seal. Apply moly-based grease to the seal lips. The seal must be pressed into place in the seal retainer until it bottoms. Ideally, this should be done by a machine shop with a press; however, you can usually do it with a large piece of wood and a hammer if you work carefully (the wood spreads the force from the hammer blows over the entire seal). Install the seal retainer, using a new gasket.

Piston rings

1   Before installing the new piston rings, the ring end gaps must be checked. It's assumed that the piston ring side clearance has been checked and verified correct (see the procedure earlier in this Chapter).

2   Lay out the piston/connecting rod assemblies and the new ring sets so the ring sets will be matched with the same piston and cylinder during the end gap measurement and engine assembly.

3   Insert the top (number one) ring into the first cylinder and square it up with the cylinder walls by pushing it in with the top of the piston (see illustration). The ring should be near the bottom of the cylinder, at the lower limit of ring travel.

4   To measure the end gap, slip feeler gauges between the ends of the ring until a gauge equal to the gap width is found (see illustration). The feeler gauge should slide between the ring ends with a slight amount of drag. The gap should be about 0.010 to 0.020 in.. If the gap is larger or smaller than specified, double-check to make sure you have the correct rings before proceeding. Too large a gap is not critical unless it exceeds about 0.038 in.

5   If the gap is too small, it must be enlarged or the ring ends may come in contact with each other during engine operation, which can cause serious damage to the engine. The end gap can be increased by filing the ring ends very carefully with a fine file. Mount the file in a vise equipped with soft jaws, slip the ring over the file with the ends con-

6.71 Installing the spacer/expander in the oil control ring groove

6.72 DO NOT use a piston ring installation tool when installing the oil ring side rails

tacting the file face and slowly move the ring to remove material from the ends. When performing this operation, file only from the outside in **(see illustration)**.

6 Repeat the procedure for each ring that will be installed in the first cylinder and for each ring in the remaining cylinders. Remember to keep rings, pistons and cylinders matched up.

7 Once the ring end gaps have been checked/corrected, the rings can be installed on the pistons.

8 The oil control ring (lowest one on the piston) is installed first. It's composed of three separate components. Slip the spacer/expander into the groove **(see illustration)**. If an anti-rotation tang is used, make sure it's inserted into the drilled hole in the ring groove. Next, install the lower side rail. Don't use a piston ring installation tool on the oil ring side rails, as they may be damaged. Instead, place one end of the side rail into the groove between the spacer/expander and the ring land, hold it firmly in place and slide a finger around the piston while pushing the rail into the groove **(see illustration)**. Next, install the upper side rail in the same manner.

9 After the three oil ring components have been installed, check to make sure that both the upper and lower side rails can be turned smoothly in the ring groove.

10 The number two (middle) ring is installed next. It's stamped with a mark which must face up, toward the top of the piston. **Note:** *Always follow the instructions printed on the ring package or box - different manufacturers may require different approaches. Do not mix up the top and middle rings, as they have different cross-sections.*

11 Use a piston ring installation tool and make sure the identification mark is facing the top of the piston, then slip the ring into the middle groove on the piston **(see illustration)**. Don't expand the ring any more than is necessary to slide it over the piston.

12 Install the number one (top) ring in the same manner. Make sure the mark is facing up. Be careful not to confuse the number one and number two rings.

13 Repeat the procedure for the remaining pistons and rings.

6.73 Installing the compression rings with a ring expander - the mark (arrow) must face up

Pistons/connecting rods

1 Before installing the piston/connecting rod assemblies, the cylinder walls must be perfectly clean, the top edge of each cylinder must be chamfered, and the crankshaft must be in place.

2 Remove the connecting rod cap from the end of the number one connecting rod. Remove the old bearing inserts and wipe the bearing surfaces of the connecting rod and cap with a clean, lint-free cloth. They must be kept spotlessly clean.

3 Clean the back side of the new upper bearing half, then lay it in place in the connecting rod. Make sure the tab on the bearing fits into the recess in the rod. Don't hammer the bearing insert into place and be very careful not to nick or gouge the bearing face. Don't lubricate the bearing at this time.

4 Clean the back side of the remaining bearing insert and install it in the rod cap. Again, make sure the tab on the bearing fits into the recess in the cap, and don't apply any lubricant. It's critically important that the mating surfaces of

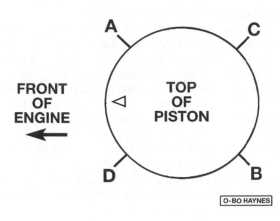

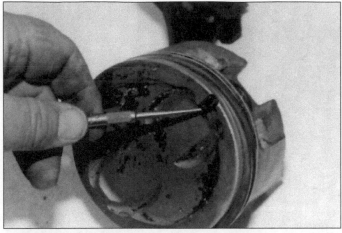

6.74 Ring end gap positions

A   *Oil rings rail gaps*
B   *Second compression ring gap*
C   *Oil ring spacer gap (position in-between marks)*
D   *Top compression ring gap*

6.75 The notch or arrow in the top of each piston must face the FRONT of the engine as the pistons are installed

the bearing and connecting rod are perfectly clean and oil free when they're assembled.

5   Position the piston ring gaps at intervals around the piston **(see illustration)**, then slip a section of plastic or rubber hose over each connecting rod cap bolt.

6   Lubricate the piston and rings with clean engine oil and attach a piston ring compressor to the piston. Leave the skirt protruding about _-inch to guide the piston into the cylinder. The rings must be compressed until they're flush with the piston.

7   Rotate the crankshaft until the number one connecting rod journal is at BDC (bottom dead center) and apply a coat of engine oil to the cylinder walls.

8   With the notch on top of the piston **(see illustration)** facing the front of the engine, gently insert the piston/connecting rod assembly into the number one cylinder bore and rest the bottom edge of the ring compressor on the engine block. Tap the top edge of the ring compressor to

make sure it's contacting the block around its entire circumference.

9   Carefully tap on the top of the piston with the end of a wooden hammer handle **(see illustration)** while guiding the end of the connecting rod into place on the crankshaft journal. The piston rings may try to pop out of the ring compressor just before entering the cylinder bore, so keep some downward pressure on the ring compressor. Work slowly, and if any resistance is felt as the piston enters the cylinder, stop immediately. Find out what's hanging up and fix it before proceeding. Do not, for any reason, force the piston into the cylinder, as you might break a ring and/or the piston!

10   Once the piston/connecting rod assembly is installed, the connecting rod bearing oil clearance must be checked before the rod cap is permanently bolted in place.

11   Cut a piece of the appropriate size Plastigage slightly shorter than the width of the connecting rod bearing and lay it in place on the number one connecting rod journal, parallel with the journal axis **(see illustration)**.

12   Clean the connecting rod cap bearing face, remove the

6.76 Drive the piston gently into the cylinder bore with the end of a wooden or plastic hammer handle

6.77 Lay the Plastigage strips on each rod bearing journal, parallel to the crankshaft centerline

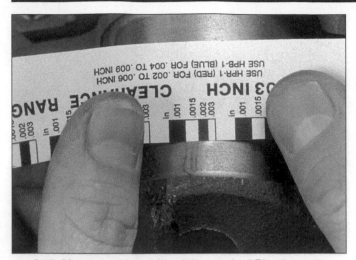

6.78 Measuring the width of the crushed Plastigage to determine the rod bearing oil clearance (be sure to use the correct scale - standard and metric ones are included)

protective hoses from the connecting rod bolts and install the rod cap. Make sure the mating mark on the cap is on the same side as the mark on the connecting rod. Install the nuts and tighten them to the correct torque, working up to it in three steps. Approximate torque figures follow: Small block - 11/32-in. bolts, 30 to 35 ft-lbs; 3/8-in. bolts, 40 to 45 ft-lbs. Big block - 3/8-in. bolts, 50 ft-lbs; 7/16-in. bolts, 67 ft-lbs. **Note:** *Use a thin-wall socket to avoid erroneous torque readings that can result if the socket is wedged between the rod cap and nut. Do not rotate the crankshaft at any time during this operation!*

13 Remove the rod cap, being very careful not to disturb the Plastigage. Compare the width of the crushed Plastigage to the scale printed on the Plastigage container to obtain the oil clearance **(see illustration)**. Generally, the clearance should be about 0.001 to 0.002 in. (for exact specifications, see the *Haynes Automotive Repair Manual* for your particular vehicle). If the clearance is not as specified, the bearing inserts may be the wrong size (which means different ones will be required). Before deciding that different inserts are needed, make sure that no dirt or oil was between the bearing inserts and the connecting rod or cap when the clearance was measured. Also, recheck the journal diameter. If the Plastigage was wider at one end than the other, the journal may be tapered.

14 Carefully scrape all traces of the Plastigage material off the rod journal and/or bearing face. Be very careful not to scratch the bearing - use your fingernail or a credit card. Make sure the bearing faces are perfectly clean, then apply a uniform layer of clean moly-base grease or engine assembly lube to both of them. You'll have to push the piston into the cylinder to expose the face of the bearing insert in the connecting rod - be sure to slip the protective hoses over the rod bolts first.

15 Slide the connecting rod back into place on the journal, remove the protective hoses from the rod cap bolts, install the rod cap and tighten the nuts to the specified torque.

Again, work up to the torque in three steps.

16 Repeat the entire procedure for the remaining piston/connecting rod assemblies. Keep the back sides of the bearing inserts and the inside of the connecting rod and cap perfectly clean when assembling them. Make sure you have the correct piston for the cylinder and that the notch on the piston faces to the front of the engine when the piston is installed. Remember, use plenty of oil to lubricate the piston before installing the ring compressor. Also, when installing the rod caps for the final time, be sure to lubricate the bearing faces using assembly lube.

17 After all the piston/connecting rod assemblies have been properly installed, rotate the crankshaft a number of times by hand to check for any obvious binding.

18 As a final step, the connecting rod endplay must be checked (see the procedure earlier in this Chapter). If it was correct before disassembly and the original crankshaft and rods were reinstalled, it should still be right. If new rods or a new crankshaft were installed, the endplay may be too small. If so, the rods will have to be removed and taken to an automotive machine shop for resizing.

Oil pump

1 Prime the oil pump prior to installation. Pour clean oil into the pickup and turn the pump shaft by hand until oil spurts out the outlet.

2 If you separate the pump from the pick-up tube, reattach it securely. A special tool is available for this purpose; however, you may be able to do it with a hammer and properly sized open-end wrench. **Note:** *An air leak at this point will cause a drop in oil pressure.*

3 Fit the driveshaft to the oil pump, using a new nylon sleeve, then fit the oil pump to the block.

4 Install the mounting bolt and tighten it securely.

Timing chain/sprockets

1 Align the keyway in the crankshaft sprocket with the Woodruff key in the end of the crankshaft. Press the sprocket onto the crankshaft with the vibration damper bolt, a large socket and some washers or tap it gently into place until it's completely seated. **Caution:** *If resistance is encountered, DO NOT hammer the sprocket onto the shaft. It may eventually move into place, but it may be cracked in the process* and fail later, causing extensive engine damage. Also, the crankshaft endplay will be disturbed.

2 Turn the crankshaft until the key is facing the two o'clock position **(see illustration)**.

3 Drape the chain over the camshaft sprocket and turn the sprocket until the timing mark faces down (6 o'clock position). Mesh the chain with the crankshaft sprocket and position the camshaft sprocket on the end of the cam **(see illustration)**. If necessary, turn the camshaft so the dowel pin fits into the sprocket hole.

4 When correctly installed, a straight line should pass through the center of the camshaft, the camshaft timing mark (in the 6 o'clock position), the crankshaft timing mark (in the 12 o'clock position) and the center of the crankshaft

6.79 Position the crankshaft with the key facing the two o'clock position, then . . .

6.80 . . . slip the chain over the crankshaft sprocket and attach the camshaft sprocket to the camshaft

(see illustration). DO NOT proceed until the valve timing is correct! Rotate the engine through two revolutions and recheck the timing marks.

5   Apply Loc-Tite to the threads and install the camshaft sprocket bolt(s). Tighten the bolt(s) to securely.

Lifters

Conventional

1   If new lifters are being installed, a new camshaft must also be installed. If a new camshaft is installed, then use new lifters as well. Never install used lifters unless the original camshaft is used and the lifters can be installed in their original locations!

Roller

2   Unlike conventional lifters, used roller lifters can be reinstalled with a new camshaft and the original camshaft can be used if new lifters are installed.

All Types

3   The original lifters, if they're being reinstalled, must be returned to their original locations. Coat them with moly-

6.81 Correctly align the timing marks (arrows)

base grease or engine assembly lube.

4   Install the lifters in the bores.

5   Install the guide plates and retainer (roller lifters only).

Notes

7 Reassembling and installing the engine

Introduction

This chapter gives the home mechanic the necessary steps and details to reassemble and install the engine. At this stage, the engine block is completely assembled (includes the cylinder heads) and all the engine components are thoroughly cleaned, painted and inspected. Make sure all the bolts and nuts have been cleaned as well.

Before installing the engine, check over the engine compartment. Be sure the engine compartment has been thoroughly cleaned and detailed (refer to Chapter - for the engine cleaning procedure). A clean engine compartment will make the maintenance much easier and enjoyable on your rebuilt engine. Also, if the engine develops a slight oil leak or coolant leak, it is much easier to detect with a clean engine compartment.

Replacing the front transmission seal

The transmission should be checked over at this point because it is much easier to access now with the engine on the bench. Transmissions commonly leak from the front seal and the fluid will drip from the bellhousing. On standard transmissions, the fluid is often mistaken for an engine rear main oil seal leak. On automatic transmissions, the fluid that leaks is sometimes black from lack of proper maintenance. Black fluid is also caused by the fluid running down the inside of the dirty bellhousing and collecting grease before it drips onto your garage floor. Now is a good time to go ahead and clean the inside of the transmission with solvent and shop rags.

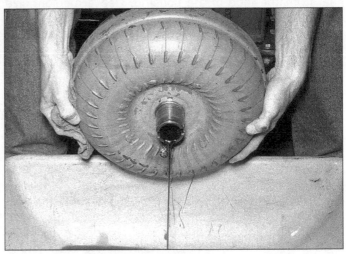

7.1 Drain the old transmission fluid out of the torque converter

Replace the front seal with a new unit. On standard transmissions, replace the seal if it is leaking or if the transmission has not been serviced for a long time (approximately 50,000 miles). On automatic transmissions, replace the seal as a matter of maintenance. Automatic transmissions require frequent servicing, so replace the seal while it is easily accessible.

On the automatic transmissions, remove the torque converter from the transmission. If the transmission fluid hasn't been changed for a while, drain the fluid into a pan or disposal barrel **(see illustration)** and add more fluid before starting the engine. Beware! The torque converter is heavy for its size. Some torque converters have a small drain bolt located on the bottom. On this type of converter it is better to drain the fluid into a pan before removing it from the transmission. Before installing the converter, it's a good

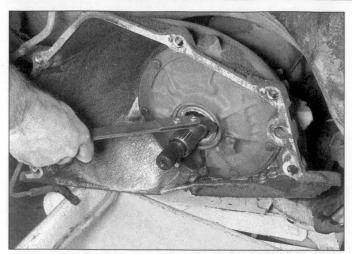

7.2 Pry the front oil seal out using a special removal tool (as shown here) or a large screwdriver

7.3 Gently tap the front seal into the bore using a large wood or metal dowel

idea to add fluid so it won't "run dry" when the engine is initially started.

After removing the torque converter, remove the front seal from the face of the housing. Use a medium size screwdriver or a seal remover tool and carefully pry the seal out **(see illustration)**. Do not damage the input shaft or the seal bore with the tool. Move the tool around to a different spot if the seal is stubborn and will not budge. Do not force the seal out with excess pressure or with a hammer and chisel. This will only cause the new seal to leak after it has been installed. Make sure the seal bore is clean and there are no gouges or deep scratches.

Prepare the front pump bore of the transmission before installing the seal. Wipe the surface clean with lacquer thinner and a paper towel. Double check for any deep scratches in the metal. On automatic transmissions, also check the bronze bushing in the front pump, below the seal. The converter drive pipe rides against this bushing, and, if it's worn, the converter and/or transmission can be damaged and the seal will be more easily damaged and prone to leaks. The bushing can be replaced with special removal and installation tools (usually available from auto parts stores). Run a bead of sealant around the outside of the seal to prevent any leakage. Use your finger to spread the sealant flat and smooth around the seal surface. It is a good idea to let the sealant "set up" (slightly harden) before installing the seal into the bore.

To install the front seal, place it squarely in the bore with the lip of the seal facing toward the torque converter. Tap the seal into place using a soft-faced hammer and a wooden dowel or punch **(see illustration)** and watch carefully that the seal goes into the bore perfectly level with the face of the housing. If you notice the seal slightly cocked, simply tap on the other side to compensate for the error. The more careful you are, the easier this job becomes! Check the torque converter before installing it into the bellhousing. Check the surface of the converter that runs up against the surface of the seal. It should be smooth and free

of scratches or burrs. Use emery cloth to smooth down any rough spots. This is a very important step for preventing transmission oil leaks. If the surface cannot be repaired, replace the torque converter with a new unit.

Lightly coat the inside of the front seal with white grease and install the torque converter. Start the converter on the transmission input shaft and rotate it back and forth while simultaneously pushing the converter. The torque converter should engage with the splines on the input shaft as well as the splines on the front pump and the stator support. You will be able to feel the converter engage the splines at each stage with a slight "click and thump". Keep rotating and pushing the torque converter if it does not engage. Don't give up! Sometimes it takes awhile. Once the torque converter is in place, position a bolt hole directly at the bottom to ease the alignment of the engine driveplate **(see illustration)**.

Look over your work and when you are satisfied with the transmission and the engine compartment - go ahead and install the components.

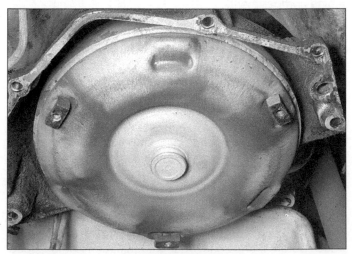

7.4 Position a torque converter bolt hole directly at the bottom of the bellhousing

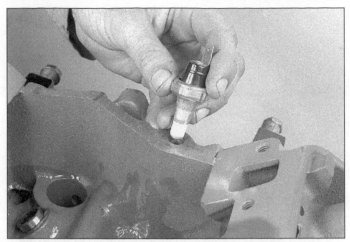

7.5 Wrap the threads on the oil pressure sending unit with thread-sealing tape before installing it into the block (small block engine shown)

Assembling the engine

Assemble the components onto the engine block. Follow the opposite order of the component removal procedure (refer to Chapter 4). Look over all the components and make sure they are cleaned and ready to be reassembled.

1 Oil pressure sending unit
2 Timing chain cover
3 Oil pan
4 Vibration damper
5 Engine mounts
6 Water pump
7 Intake manifold
8 Thermostat
9 Fuel pump
10 Distributor
11 Valve covers
12 Carburetor/throttle body
13 Exhaust manifolds
14 Accessories (alternator/air pump/ air conditioner compressor/power steering pump)
16 Flywheel/driveplate

Oil pressure sending unit

Install the oil pressure sending unit into the block. Wrap the threads of the switch with Teflon tape **(see illustration)**. Use a deep socket or an open-end wrench to install the unit without damaging the electrical connector. If you suspect the switch of any oil leaks because of a cracked insulator, be sure to replace the part with a new one!

Timing chain cover

Before installing the timing chain cover, replace the front crankshaft oil seal. There should be a new seal with your engine gasket kit. Use a punch or screwdriver and hammer to drive the seal out of the cover. Support the cover as close to the seal bore as possible. Be careful not

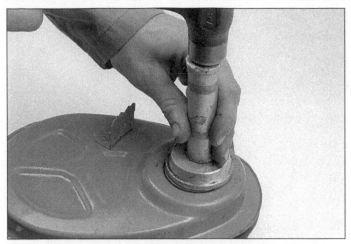

7.6 Use a special seal driver tool to install the timing cover front seal

to distort the cover or scratch the wall of the seal bore. If the engine has accumulated a lot of miles, apply penetrating oil to the seal-to-cover joint on each side and allow it to soak in before attempting to drive the seal out.

After the front seal is removed, clean the bore to remove any old seal material and corrosion. Support the cover on blocks of wood and position the new seal in the bore with the open end of the seal facing IN. A small amount of oil applied to the outer edge of the new seal will make installation easier - don't overdo it!

Drive the seal into the bore with a seal driver tool and hammer **(see illustration)** until it's completely seated. Select a driver that's the same outside diameter as the seal.

After the front seal is installed correctly, reassemble the timing chain cover onto the front of the engine block. Check over the gasket surfaces of the engine block and the cover for any left-over gasket material. Look over the timing chain cover for any cracks, warpage or other damage and replace the cover, if necessary. Apply a thin coat of contact cement to the gasket surface of the timing chain cover. Allow the

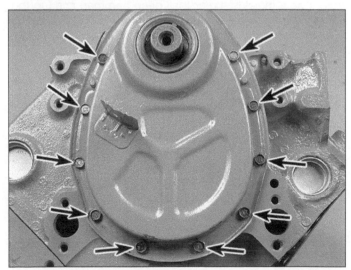

7.7 Be sure to install the timing cover bolts in their proper holes (arrows)

7.8 Use a contact cement to seal the oil pan gasket to the engine block

7.9 Use a small screwdriver to tuck the rubber seal into the cover and block

contact cement to "set up" before you attach the gasket. Wait until the sealant starts to become slightly stiff or "tacky". Position the gasket onto the timing chain cover. Apply a light coat of sealant to the cover mating surface on the engine block. Install the cover, placing the correct length bolts into the proper holes **(see illustration)**. Torque them to about 10 to 12 ft-lbs. Consult your vehicle's *Haynes Automotive Repair Manual* for exact specifications and additional information.

Oil pan

Install a new gasket onto the engine block and place the oil pan into position. If you have the engine on an engine stand, rotate the engine 180-degrees or until the bottom of the engine faces up. This is by far the easiest method. Make sure the oil pump and pick-up tube are in place and the gasket surfaces do not have any leftover gasket material.

If the oil pan lip is dented or deformed and is damaged to the point that it will leak, repair the surface. Place the oil pan on a flat metal object such as the vise on the work table. Use a metal hammer (ball peen hammer) and tap the edges until they are flat again. Turn the oil pan upside down

and lay it on the work-table or another level area and check to make sure the oil pan is straight. Keep working the damaged areas until they are flat once again. If the oil pan is severely damaged, replace it with a new unit.

With the engine block upside down, apply a thin layer of contact cement to the oil pan side rails on the block **(see illustration)**. Use your finger and spread the sealant over the block surface until it is uniform and thin. Do not "cake up" the block by applying too much! Install the gasket(s) onto the block. Install the rubber gaskets over the front cover and rear main cap seal, if equipped **(see illustration)**. **Note:** *Chevrolet engines are equipped with either a "thin" or "thick" rubber gasket on the front* **(see illustration)**. *Install each one and temporarily place the oil pan over each gasket to determine which seal fits best.* Notch the gaskets together. Apply a dab of RTV sealant into the corners over the notches and work the sealant into the grooves **(see illustration)**. Be a little more generous on this step. Make a small mound of sealant in each corner to allow for spreading when the oil pan is installed. Allow the sealant to "set up" for approximately 5 to 10 minutes depending on the air temperature.

Install the oil pan and the bolts and nuts. Tighten them

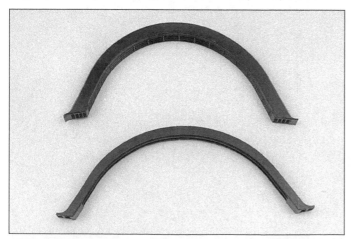

7.10 Before applying any RTV sealer, try each rubber gasket to determine the correct size for your engine

7.11 Apply a thin bead of silicone sealer into the corners where the cork gasket and the rubber gasket notch together

7.12 Apply white grease to the inside and outside of the vibration damper to ease installation and prevent the formation of rust

7.13 Use a special tool to install the vibration damper - the box-end wrench holds the main shaft or spline stationary while the adjustable wrench rotates the large nut that pushes the damper onto the crankshaft

in a series of three or four rotations to avoid pinching the oil pan gasket on one side. Use a light torque to avoid meshing the gasket and causing an oil leak. Torque the oil pan bolts to 7 to 9 ft-lbs. on 1\4 in. bolts and 9 to 11 ft-lbs. on 5\16 in. bolts. Consult your vehicle's *Haynes Automotive Repair Manual* for exact specifications and additional information.

Vibration damper

Install the vibration damper onto the crankshaft. Be sure the damper is not bent or damaged to the point of causing unnecessary vibration. If necessary, replace the vibration damper with a new unit. Coat the inside of the damper with anti-seize compound or grease **(see illustration)**. This will help the damper to slide over the crankshaft and prevent the damper from rusting onto the crankshaft miles down the road. Use a special vibration damper installation tool and screw the damper flush with the crankshaft **(see illustration)**. Use a box-end wrench to hold the main shaft stationary. Watch carefully that the damper goes in level and without binding. **Note:** *Do not interchange the vibration dampers from one size engine to a different size engine or you will increase engine noise and possibly damage the crankshaft.*

Torque the damper to 70 to 90 ft-lbs. Consult your vehicle's *Haynes Automotive Repair Manual* for exact specifications and additional information.

Engine mounts

Install the engine mounts to the block. If the mounts are worn or damaged, be sure to replace them with new units - it's a lot easier to do now than when the engine is installed.

Water pump

Install the water pump on the front of the engine block. Apply a thin layer of RTV sealant to the mating surfaces of the water pump and the timing chain cover. Allow the sealant to "set up." Position the water pump gasket onto the water pump (the gasket should stick to the surface of

the water pump). Lightly oil all the bolts and attach the pump onto the engine block. Arrange the correct length bolts into the proper holes. Use the diagram that you drew as a guide (refer to Chapter 4).

Torque the water pump bolts; the torque ranges from 18 to 30 ft. lbs. Consult your vehicle's *Haynes Automotive Repair Manual* for exact specifications and additional information.

Intake manifold

Note: *The mating surfaces of the cylinder heads, block and manifold must be perfectly clean when the manifold is installed. If you have not "hot tanked" the intake manifold, gasket removal solvents in aerosol cans are available at most auto parts stores and may be helpful when removing old gasket material that's stuck to the heads and manifold. Aggressive scraping can cause damage! Be sure to follow the directions printed on the container.*

If the manifold was disassembled, reassemble it **(see**

7.14 Remove the intake manifold bolts (arrows) - 350 engine shown

7.15 Seal the circumference of the water ports with contact cement

7.16 Be sure THIS SIDE UP is clearly visible on the intake manifold gasket (arrow)

7.17 Seal the gasket-to-intake manifold side with contact cement

7.18 If the gasket set is not equipped with the rubber gaskets, run a bead of silicone sealant across the rim (front and rear) of the engine block

illustration). Use Teflon tape on the temperature sending unit threads. Use a new EGR valve gasket. Use a gasket scraper to remove all traces of sealant and old gasket material, then clean the mating surfaces with lacquer thinner or acetone. If there's old sealant or oil on the mating surfaces when the manifold is installed, oil or vacuum leaks may develop.

The intake manifold gaskets come in a series of different sizes and materials. The gaskets that seal the cylinder heads to the intake manifold are usually metal and the gaskets that seal the front and rear sections of the block are usually rubber. **Note:** *The gasket kit is equipped with a pair of metal restrictors that are positioned in the center port of the intake manifold gasket. Refer to the gasket instructions to find out if your particular year and engine are required to have them installed.*

Apply a 1/8-inch wide bead of RTV sealant to the four corners where the manifold, block and heads converge. **Note:** Be aware that the sealant sets up very quickly (faster in warm temperatures). Do not take long to install and

tighten the manifold once the sealant is applied, or leaks may occur. Apply a small dab of contact adhesive to the manifold gasket mating surface on each cylinder head **(see illustration)**. Position the gaskets on the cylinder heads **(see illustration)**. The upper side of each gasket will have a TOP or THIS SIDE UP label stamped into it to ensure correct installation. Also, apply a small amount of contact adhesive to the intake manifold gaskets **(see illustration)** around the water ports. Position the end seals on the block, then apply a 1/8-inch wide bead of RTV sealant to the four points where the end seals meet the heads **(see illustration)**.

Make sure all intake port openings, coolant passage holes and bolt holes are aligned correctly.

Carefully lower the manifold in place once the sealant has "set up" **(see illustration)**. **Caution:** *Don't disturb the gaskets and don't move the manifold fore-and-aft after it contacts the seals on the block. Make sure the end seals haven't been disturbed.*

7.19 Wait several minutes and make sure the silicone has "set-up" (slightly hardened) and lay the intake manifold onto the engine block

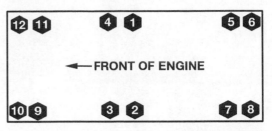

24065-2A-6.16B HAYNES

7.20a Intake manifold bolt torque sequence (small-block engine) - tighten the bolts smoothly and evenly, in this sequence, to 30 ft-lbs

Install the bolts and tighten them to the correct torque. Consult your vehicle's *Haynes Automotive Repair Manual* for exact specifications and additional information. Follow the recommended sequence **(see illustrations)**. Work up to the final torque in three steps.

Recheck the mounting bolt torque and remove the guide pins.

Thermostat

Install the thermostat into the intake manifold and position the housing **(see illustration)**. Check the gasket mating surfaces of the intake manifold and the thermostat housing for any excess gasket material. Use a gasket scraper and remove anything left over. Spray a small amount of lacquer thinner onto the surfaces and wipe them down. Look again and then run your finger over the area. A small chunk of material can cause a small but irritating coolant leak running from the thermostat housing down the front of the engine onto the floor.

Apply a small amount of RTV sealant to the gasket mating surfaces on the thermostat housing and the intake manifold. Use your finger and spread the sealant thinly and

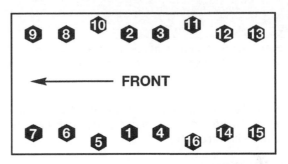

24064-2A-6.15b HAYNES

7.20b Intake manifold bolt torque sequence (big-block engine) - tighten the bolts smoothly and evenly, in this sequence, to 30 ft-lbs

evenly. Wait several minutes and allow the sealant to "set up" before you install the housing. Position a new gasket onto the thermostat housing and install the bolts. Place the housing on the engine and tighten the bolts securely.

Fuel pump

Install the fuel pump and gasket onto the engine block. Be sure to coat the pushrod rod with grease **(see illustration)**. Tighten the bolts securely. **Note:** *On small-block*

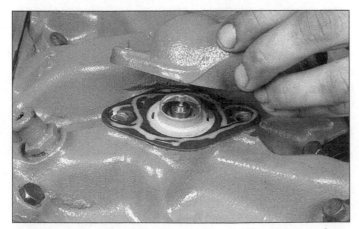

7.21 Apply contact cement to the thermostat gasket surface

7.22 Apply grease to the fuel pump pushrod to hold it in place inside the housing while the fuel pump is being installed

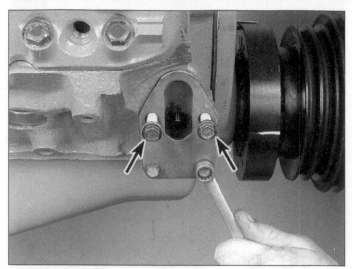

7.23 Install the fuel pump plate and gasket with the fuel pump bolts (arrows) loosely in place in order to keep the plate in alignment

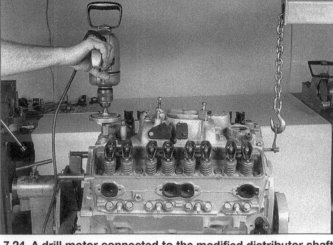

7.24 A drill motor connected to the modified distributor shaft drives the oil pump - make sure its turns clockwise as viewed from above

engines, install the upper mounting bolts into the fuel pump plate when tightening the lower bolts to keep the plate aligned (see illustration).

Distributor

Priming the oil pump

Before installing the distributor, it is necessary to prime the oil pump. The oil pump driveshaft is easily accessible at this point. The oil pan should be secure and ready to be filled with oil. Pour about two quarts of oil through the distributor hole in the block and give it to completely drain into the pan.

The distributor shaft normally drives the oil pump, but, for purposes of priming the bearings, camshaft and crankshaft, connect a special tool onto the oil pump drive-

shaft. A modified small-block or big-block Chevrolet distributor will be needed for this procedure - a junkyard should be able to supply one for a reasonable price. In order to function as a pre-oil tool, the distributor must have the gear on the lower end of the shaft ground off (see illustration) and, if equipped, the advance weights on the upper end of the shaft removed.

Install the pre-oil distributor in place of the original distributor and make sure the lower end of the shaft mates with the upper end of the oil pump driveshaft. Turn the distributor shaft until they are aligned and the distributor body seats on the block. Install the distributor hold-down clamp and bolt.

Mount the upper end of the shaft in the chuck of an electric drill and use the drill to turn the pre-oil distributor shaft, which will drive the oil pump and circulate the oil throughout the engine (see illustration). **Note:** *The drill must turn in a clockwise direction.*

It may take two or three minutes, but oil should start to flow out of all the rocker arm holes, indicating that the oil pump is working properly. Let the oil circulate for several seconds, then shut off the drill motor.

Once you see the oil, look carefully that it flows evenly and uniformly over each rocker. If necessary, install an oil pressure gauge into the oil pressure sending unit hole and obtain a pressure reading. Once the oil is flowing properly, remove the oil priming tool.

Locating Top Dead Center (TDC) for the number one piston

Before you can actually insert the distributor into the block, you must have the camshaft and crankshaft timed at number one cylinder Top Dead Center (TDC). Top Dead center (TDC) is the highest point in the cylinder that each piston reaches as it travels up-and-down when the crankshaft turns. Each piston reaches TDC on the com-

7.25 Oil, assembly lube, or grease will begin to flow from all of the rocker arm holes if the oil pump and lubrication system are functioning properly

7.26 Rotate the vibration damper until 0 degrees is on the pointer (A) and the intake (B) and exhaust (C) valves are both closed (springs not compressed and rocker arms loose) - the engine is on number 1 TDC

7.27 If the exhaust valve (arrow) is open, you must rotate the crankshaft through another revolution and again align the marks

pression stroke and again on the exhaust stroke, but TDC generally refers to piston position on the compression stroke. The timing marks on the vibration damper installed on the front of the crankshaft are referenced to the number one piston at TDC on the compression stroke. Number one cylinder TDC is the position that is used to install the distributor, start the valve adjustment procedure and detect engine timing with a stroboscopic timing light.

Be aware of the fact that the damper and crankshaft will rotate two complete revolutions before all eight cylinders complete their cycle (firing order). This will cause the O-degree mark on the vibration damper to pass the timing pointer twice; one time the number one cylinder will be at TDC on the compression stroke, and one time it will be at TDC on the exhaust stroke.

In order to bring any piston to TDC, the crankshaft must be turned. When looking at the front of the engine, normal crankshaft rotation is clockwise. The preferred method is to turn the crankshaft with a large socket and breaker bar attached to the vibration damper bolt that is threaded into the front of the crankshaft. Refer to the distributor removal procedure in Chapter 4.

Turn the crankshaft until the zero or groove on the vibration damper is aligned with the pointer or TDC mark **(see illustration)**.

To get the piston to TDC on the compression stroke, you must first determine if the valves on the number one cylinder are in the closed position (compression stroke) or in the open position (exhaust stroke). If the valves are closed, the rocker arms should be level and the valve springs at resting height (not compressed). If the piston is at TDC on the exhaust stroke, one of the valve springs (exhaust) will be compressed **(see illustration)**. If one of the valves is open, the number six (companion cylinder) NOT number one is at TDC on the compression stroke. This is

very important,. Rotate the crankshaft pulley 360-degrees and check again. Both valves on the number one cylinder SHOULD BE CLOSED when the timing is set at number one TDC.

After the number one piston has been positioned at TDC on the compression stroke, the cylinder firing position for any of the remaining cylinders can be located by turning the crankshaft and following the firing order, which should be stamped into the intake manifold. (1-8-4-3-6-5-7-2).
Note: *The terminal numbers are often marked on the spark plug wires near the distributor. It is important to remember that the valves for each particular cylinder must be CLOSED when the cylinder is ready to ignite on its compression stroke. This can be determined from the position of the rocker arms for each particular cylinder.*

Installing the distributor

Now that the engine is set up for number one TDC and the oil pump is primed and ready to go, install the distributor into the engine block. Normally the distributor will slide right into place, but don't get frustrated if it gets stuck and takes a few tries or doesn't line up right the first time. The biggest hindrance is the oil pump driveshaft. The oil pump driveshaft must be aligned just right to accommodate the distributor shaft.

As you lower the distributor into place, notice the white marks you painted on the rim of the distributor and on the engine (see Chapter 4). These marks must be aligned. If you no longer have marks (say, for in-stance, you're installing a new distributor), set the distributor cap into place, find the number one spark plug terminal and point the rotor toward that spot. Try to keep the distributor positioned so the vacuum advance unit will not interfere with anything when it's rotated a couple of inches in either direction

Since the distributor drive gears are angle-cut, you'll have to start dropping the distributor into place with the

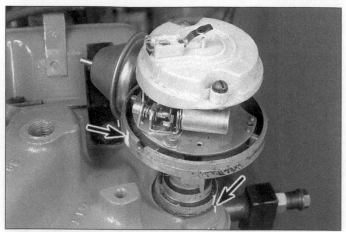

7.28 When installing the distributor, align the rotor with the two white paint marks you made during disassembly - one on the rim of the distributor body (arrow) and one on the engine block (arrow) - this will assure the engine is timed so it will fire right away when it's started

7.29 Rotate the pushrod while slowly tightening the rocker arm nut to find the exact point when the lifter starts to collapse. This is evident when the pushrod no longer spins freely (small block shown)

rotor slightly to the side of the mark **(see illustration)**; the rotor will move as the distributor is dropped down. When the distributor base is flush with the engine and the rotor is pointing to the paint marks, you've done the job correctly.

Adjusting the valves

You must adjust each valve in its fully closed position. The adjustment method depends on which engine you have. Engines with mechanical lifters (very rare) are adjusted to produce a certain amount of lash or clearance between the valve tip and the rocker arm. If you've got an engine that uses hydraulic lifters, the adjustment is for the initial lifter setting.

Rotate the crankshaft until the pointer aligns with number one TDC (see the TDC locating procedure above). Both valves of a cylinder are closed when the piston is at TDC on its compression stroke, so position each cylinder at TDC before adjusting its valves (see the TDC locating procedure above).

Hydraulic lifters

Run the adjusting nut down its stud until the slack is removed from the rocker and pushrod. Verify the slack is gone - but the lifter plunger has yet to collapse - by rotating the pushrod (when there's no slack, it won't rotate anymore). When the slack is taken up, give the adjusting nut another full-turn **(see illustration)**. That's it - the valve is adjusted! Proceed to the next valve and repeat this procedure for the remaining cylinders. The object here is to collapse the lifter plunger to the midpoint of its travel. If the lifter is fully primed, it may not collapse just yet. It'll take some time before it bleeds down.

Mechanical lifters

With the piston at TDC, adjust the lash, which is the clearance between the rocker arm and the valve tip. Insert the feeler gauge between the rocker arm and valve tip, wig-

gle the rocker arm to make sure there's no slack in the valve train and adjust the nut so there's some drag on the feeler gauge as you slide it between the valve tip and the rocker arm (it should feel like the gauge is being drawn across a magnet). With the (cold) lash set, loosely install the jam nuts on the adjusting nuts. You'll have to adjust the valve to a different specification after the engine is warmed up (hot lash). You can tighten the jam nuts after hot lashing. For lash specifications, check the *Haynes Automotive Repair Manual* for your particular vehicle or check with the camshaft manufacturer.

Valve covers

Install the valve-cover gaskets on the valve covers. Make sure the gasket tabs engage in the slots on the valve cover (if equipped). To ensure that these gaskets stay in place while you install the valve covers, use contact cement between the gasket and the cover. Then, if necessary, you can remove the valve covers without destroying the gaskets. This is especially helpful on engines with mechanical lifters, since the valve covers have to come off later for a hot lash adjustment.

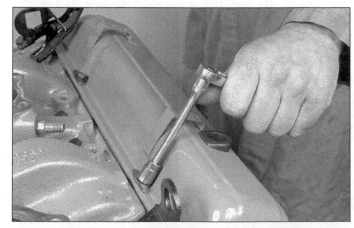

7.30 Do not overtighten the valve cover bolts

7.31 Use only new carburetor gaskets when installing the carburetor - make sure the gasket mating surfaces are completely clean

Once the gaskets are in place, install the valve covers **(see illustration)**. Don't overtighten the bolts. The only thing you'll succeed in doing is bending the valve-cover flanges, which will increase the likelihood of an oil leak. Gradually tighten the bolts to compress the cork gasket, then torque the bolts very lightly (to about 5 ft-lbs.).

Carburetor/throttle body

Install the carburetor or the throttle body onto the intake manifold. Inspect the surfaces of the carburetor baseplate and the underside of the carburetor itself. Remove any pieces of old gasket material using a gasket scraper. Install new gaskets on the top and bottom of the carburetor baseplate - if equipped - **(see illustration)**. Install the carburetor or throttle body nuts/bolts tighten them securely. Consult your vehicle's *Haynes Automotive Repair Manual* for torque specifications and additional information.

Exhaust manifolds

Install the exhaust manifolds onto the cylinder heads. Check all gasket surfaces carefully for any leftover pieces of gasket material. Install new exhaust gaskets and torque the bolts to about 18 to 24 ft-lbs. Consult your vehicle's *Haynes Automotive Repair Manual* for exact specifications and additional information. **Note:** *Many engines come from the factory new with no gaskets. This is because of precise machining tolerances. The heating/cooling cycle of the manifolds and cylinder heads once in service causes warpage that makes it impossible to reattach the manifolds without gaskets.*

Accessories (alternator/air pump/air conditioning compressor/power steering pump)

Install the accessories onto the engine. Refer to Chapter - for the removal procedure and reverse the steps. Consult your diagrams to help locate all the correct size bolts and brackets **(see illustration)**. **Note:** *When in- stalling the air injection pump bracket, use RTV sealer on the bolt threads to prevent oil leaks.*

7.32 Use diagrams or photos of the accessories to help the installation (1973 350 engine shown)

7.33 Use thread locking compound on the threads of the flywheel/driveplate bolts

Flywheel/driveplate

Note: *If you have a manual transmission, inspect the clutch, as described below, before you install the flywheel.*

The final components to be assembled onto the long block are the flywheel and clutch assembly on a manual transmission or the driveplate on the automatic transmission. Attach the chain and hoist to the engine (refer to Chapter 4) and raise the engine with the stand slightly off the ground. Remove the stand from the engine harness and disconnect the engine harness from the engine block. Now you have access to the rear of the engine. Lower the engine onto a wood pallet and/or onto the floor of the shop just enough to take the weight of the engine off the hoist. Don't bother to disconnect the chain assembly.

Align the paint marks on the crankshaft and flywheel/driveplate to verify the correct alignment (see Chapter 4). Install the flywheel/driveplate bolts **(see illustration)**. Torque the bolts to about 75 to 85 ft-lbs. Consult your vehicle's *Haynes Automotive Repair Manual* for exact specifications and additional information.

7.34 Use a clutch alignment tool to center the clutch disc, then tighten the pressure plate bolts

7.35 Make sure the clutch disc is installed with the marked face against the flywheel

On models with manual transmissions, clean the flywheel and pressure plate machined surfaces with lacquer thinner or alcohol. It's important that no oil or grease is on these surfaces or the lining of the clutch disc. Handle the parts only with clean hands. Position the clutch disc and pressure plate against the flywheel with the disc held in place with an alignment tool **(see illustration)**. Make sure it's installed properly (most replacement clutch plates will be marked "flywheel side" or something similar. If not marked, install the clutch disc with the damper springs toward the transmission) **(see illustration)**. Tighten the pressure plate-to-flywheel bolts only finger tight, working around the pressure plate. Center the clutch disc by ensuring the alignment tool extends through the splined hub and into the pilot bearing in the crankshaft. Wiggle the tool up, down or side-to-side as needed to bottom the tool in the pilot bearing. Tighten the pressure plate-to-flywheel bolts a little at a time, working in a criss-cross pattern to prevent distorting the cover. After all of the bolts are snug, tighten them securely. Check your vehicle's *Haynes Automotive Repair Manual* for exact specifications and additional information. Remove the alignment tool.

Inspecting the clutch/flywheel

Warning: Dust produced by clutch wear and deposited on clutch components may contain asbestos, which is hazardous to your health. DO NOT blow it out with compressed air and DO NOT inhale it. DO NOT use gasoline or petroleum-based solvents to remove the dust. Brake system cleaner should be used to flush the dust into a drain pan. After the clutch components are wiped clean with a rag, dispose of the contaminated rags and cleaner in a covered, marked container.

If you have a standard transmission, inspect the clutch assembly (pressure plate, disc and release bearing) and replace any worn parts **(see illustration)**. If the clutch slipped before the engine was removed, the chances are the clutch assembly should be replaced with a new unit. If the clutch was adequate before the engine was rebuilt, the

clutch assembly should be carefully checked to determine how many miles the clutch has left before it will have to be replaced. Then, you can decide if you want to spend the extra MONEY NOW and replace the clutch assembly or wait to spend the extra LABOR LATER removing the trans-

NORMAL FINGER WEAR

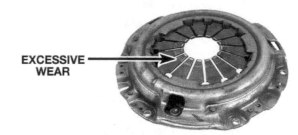

EXCESSIVE WEAR ⟶

EXCESSIVE FINGER WEAR

BROKEN OR BENT FINGERS

7.36 Replace the pressure plate if excessive wear is noted

7.37 This clutch is badly worn out - the disc has worn down to the rivets . . .

7.38 . . . damaging the pressure plate, which must now be replaced

mission. At any rate, the clutch assembly should be examined in order to make an intelligent choice.

Inspect the clutch disc for thickness and general wear conditions. If the disc has been used excessively, the disc material will be worn down exposing the rivets **(see illustration)**. Once this occurs, the metal rivets will contact the pressure plate **(see illustration)** and the flywheel and will score or gouge the surfaces. This will require a new clutch assembly and a flywheel resurfacing (the surface cut or milled level) by a qualified machine shop. If the machinist determines the flywheel thickness not adequate for machining, replace the flywheel with a new unit. If the disc looks OK, measure it to verify its condition. Clamp the disc in a vise but be careful not to damage the facings or allow grease or oil on them. Squeeze the disc just enough to compress the wave spring into a flat position. Measure the disc thickness as close as possible to the clamp using a micrometer. The thickness of a new clutch disc is about 0.330-inch which allows for 0.050-inch wear. The rivets will be level with the face of the disc at 0.280-inch. You should replace the disc at this point. Anything close to this measurement is just a matter of time. Use your own judgement based on your budget.

Inspect the surfaces of the flywheel and the pressure plate. A good general rule is: if the disc is completely worn out, replace the pressure plate assembly; if the disc is OK, inspect the pressure plate carefully and replace only the necessary components. Simply look at the surface and check for unusual wear. If the friction surface is bright and shiny, and free of any grooves or gouges, the surface is OK and can be reused. If the surface has black and blue marks (hot spots), the pressure plate is damaged and must be replaced with a new unit and the flywheel must be resurfaced by a qualified machine shop. Also check the surface to make sure it is level. Lay a straightedge or a heavy-steel rule over the pressure plate surface. If there is an obvious warp, replace the pressure plate with a new unit. Also, if the surface is cracked or grooved, replace the pressure plate with a new unit. If the same conditions exist on the surface of the flywheel, take caution and inspect the flywheel care-

fully. Hot spots and small grooves (and even some surface cracks) can be removed by a qualified machinist, but cracks must be inspected very closely. If crack(s) have developed into the flywheel surface and continue growing, the flywheel must be replaced with a new unit. Have an automotive machine shop check the cracks. A cracked flywheel can easily explode, flying apart with the force of a hand grenade!

If the flywheel is determined to be OK, it should be cleaned. Use fine grit sandpaper to rough up the surface. Keep working the surface until the glazed, shiny surface has been thoroughly scratched and dulled. This will give the surface a tacky or gritty texture. Follow the sanding by cleaning the surface with a non-petroleum solvent such as lacquer thinner or alcohol. This will remove any oil deposits. Clean the surface of the pressure plate also.

If hot spots, warpage or cracks are minimal and the overall condition of the pressure plate does not look too bad, you might consider the price of a rebuilt pressure plate versus the price of a new one. Check the guarantees and prices for each type (new and rebuilt) and make your decision based on the facts. If there is any sign of trouble with the clutch components, replace them with new or rebuilt units.

Remove the clutch release lever from the ball-stud by sliding it off the ball-stud (if equipped) and pulling it through into the bellhousing. Then remove the bearing from the lever. Check the release bearing for wear and replace it with a new unit, if necessary. Place the bearing on a firm surface. With the palm of your hand, press against the surface that normally contacts the pressure plate fingers and rotate your palm back and forth. It should rotate smoothly with no binding, roughness or noise. If not, replace it. Since the bearing isn't normally very expensive, and since it's a big pain to remove the transmission just to replace the bearing, you should replace the bearing as long as you've got it out - they do fail in service!

Lightly lubricate the clutch lever crown and spring retention crown where they contact the bearing with high-temperature grease. Fill the inner groove of the bearing with

7.39 Pack the cavity behind the pilot bearing with heavy grease and force it out hydraulically with a rod slightly smaller than the bore in the bearing - when the hammer strikes the rod, the bearing will pop out of the crankshaft

Installing the engine

Finally, the moment has come! Your engine components are all properly installed onto your engine block and it is time to raise it up and install it into your vehicle. Double check that the chain and lifting lugs are secure and ready to go. If you are working with an automatic transmission, make sure the marked hole on the driveplate is at the bottom. The marked stud on the torque converter should be positioned at the bottom also. Raise the engine into position. Watch out for the emblem or ornament on the body as the engine glides over the front end. If you have one, use an adjustable engine sling (refer to Chapter 4) to change the angle of the engine. It is better to slightly tilt the engine toward the transmission as it is lowered.

Position a floorjack under the transmission and raise it up and remove the bar that straddles the engine compartment. Disconnect the bolts and anything else that was supporting the transmission and exhaust pipes.

On manual transmissions, lightly oil the crankshaft pilot bearing, dab a small amount of grease onto the transmission input shaft and put the transmission in second gear. Look over the engine compartment to make sure nothing will interfere with the installation of the engine.

Slowly lower the engine into the engine compartment. Raise the front of the transmission as high as possible to ease the installation and to help clear the engine mounts.

On manual transmissions, center the clutch assembly with the input shaft until they are positioned (leveled) on the same plane. Do not let the weight of the engine rest on the input shaft as the engine is being lowered. Watch carefully at all times that the engine does not "hang up" or get jarred into the transmission. Work the engine into the transmission bellhousing as the engine is slowly lowered. The installation will go much smoother if the engine is lowered an inch at a time and if the engine and transmission alignment is constantly checked. Once the clutch disc and input shaft contact each other, place a socket and wrench onto the front pulley and rotate the crankshaft back and forth until the splines "lock in". After they engage, place the transmission back into Neutral to allow the engine to align with the dowel pins in the transmission bellhousing. Carefully work the engine into the transmission bellhousing until the engine and transmission mate together. The engine and transmission will literally "click" when the two come together. Install two of the bolts into the bellhousing.

On automatic transmissions, the torque converter studs and the dowel pins must engage almost simultaneously. Push the engine into the transmission but do not force it. Here again, the angle of the engine in relation to the transmission is extremely important for a smooth installation. If the engine gets "hung up" check the alignment of the torque converter and driveplate. Also, check the engine mounts and block to make sure they are not locked prematurely.

Once the engine and transmission is mated and the bellhousing bolts are tight, gently lower the engine onto the

the same grease. Attach the release bearing to the clutch lever. Lubricate the clutch release lever ball-stud socket with high-temperature grease and push the lever onto the ball stud until it's firmly seated. Apply a light coat of high-temperature grease to the face of the release bearing, where it contacts the pressure plate diaphragm fingers.

Inspect the pilot bearing. If the clutch assembly is replaced, install a new pilot bearing also. The pilot bearing is a needle roller type bearing which is pressed into the rear of the crankshaft. It is greased at the factory and does not require additional lubrication. Its primary purpose is to support the front of the transmission input shaft. The pilot bearing should be inspected whenever the clutch components are removed from the engine. Due to its inaccessibility, if you are in doubt as to its condition, replace it with a new one. Inspect for any excessive wear, scoring, lack of grease, dryness or obvious damage. If any of these conditions are noted, the bearing should be replaced. A flashlight will be helpful to direct light into the recess.

You can remove the bearing with a special puller and slide hammer, but an alternative method also works very well **(see illustration)**. Find a solid steel bar which is slightly smaller in diameter than the bearing. Alternatives to a solid bar would be a wood dowel or a socket with a bolt fixed in place to make it solid. Check the bar for fit - it should just slip into the bearing with very little clearance. Pack the bearing and the area behind it (in the crankshaft recess) with heavy grease. Pack it tightly to eliminate as much air as possible. Insert the bar into the bearing bore and strike the bar sharply with a hammer which will force the grease to the back side of the bearing. Clean all grease from the crankshaft recess.

To install the new bearing, lightly lubricate the outside surface with lithium-based grease, then drive it into the recess with a hammer and socket slightly smaller than the outside diameter of the bearing. The seal must face out.

motor mounts or onto the frame. Lower the engine until it contacts the surface and align the engine mounts. Do not lower the engine all the way until the mounts are aligned. Be patient with this step. What came apart easily is very tough to put together. Use a crowbar or use an assistant help position the engine into the exact correct position. It is a great help to have one person work the left engine mount and the other person work the right engine mount. Install the through bolts (if equipped) or the mounting nuts and tighten them securely.

Once the engine mounts are attached and the engine bolted to the transmission, remove the engine hoist and the engine chain from the work area. Be sure to replace any manifold bolts with the original after removing the engine chain and lugs.

Check to make sure the jackstands under the vehicle are secure and the work area is safe. Roll under the vehicle on your creeper and install any remaining bolts into the bell-housing or engine mounts. Check each one carefully and tighten them securely. Check your vehicle's *Haynes Automotive Repair Manual* for exact specifications and additional information.

Install the remaining components by reversing the procedures in Chapter 4. Use anti-seize compound on the exhaust system fasteners so they won't get stuck together and be impossible to remove later. When routing the spark plug wires, use plastic clips, keep them away from hot components and try not to have two wires run next to each other for long distances; this can cause induced-current cross-fire **(see illustration)**. Fill the cooling system with a 50/50 mixture of coolant and water. Fill the system slowly, allowing it to purge trapped air. It may take 1/2 hour or more for all the air to be expelled. Add enough oil to bring the level up to normal. If the vehicle is equipped with an automatic transmission and the torque converter was removed, the fluid level may not be correct.

Starting the engine

Note: *Since you must monitor gauges, listen for noises and look for leaks - all at once - when the engine fires, it helps to have an assistant on hand.*

The moment has finally arrived! The engine is installed and the cooling, starting and fuel systems are all in order and the engine is ready to be started. Look over the engine compartment one last time. Remove any tools, rolls of tape or loose clamps and bolts from the engine compartment. Check for potential (or active) fuel, oil and coolant leaks. Make sure everything is mounted securely, including the distributor, radiator cap, oil filter, oil cap and fuel filter. Check all the electrical connections to make sure they're correct and tight. Make sure the battery is fully charged. Start the engine.

If the engine cranks over but does not fire, stop for a moment and check to make sure the coil wire is positioned in the distributor cap and in the coil. Also, check the carburetor for fuel. If the fuel is being pumped to the carburetor

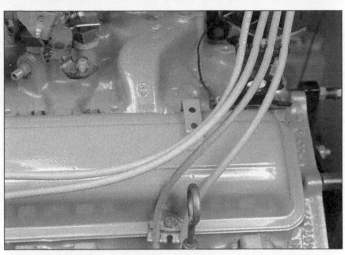

7.40 Use spark plug wire clips to hold them in place and keep them organized - route the wires away from components that get hot when the engine is running

you will smell it once you get close to the carburetor. Add a small amount (2 or 3 tablespoons) to the carburetor throat to prime it. If the engine starts momentarily and then dies, suspect the fuel pump or fuel system. If the engine continues to not even spark or fire, suspect the electrical system or the ignition system.

If the engine sputters and backfires, check the firing order. If it's OK, the ignition timing may be off. Try loosening the distributor hold-down and slowly rotating the distributor in either direction while an assistant cranks the engine. If you've done any work on the distributor, the point gap may not be correct.

If the engine starts up and the lifters clatter, do not be alarmed. Wait two or three minutes and the noise should dissipate as the oil is "pumped up" to the lifters. Allow the engine to run at 1200 to 1500 rpm to keep the oil pressure up and the fan blowing cool air over the radiator. Be aware of the rpm level; idle speed (750 to 900 rpm) is too slow while excessive rpm (2000) will overwork the new engine parts. Keep an eye on the oil pressure gauge or light and the water temperature gauge to make sure everything stays normal. Also keep an eye on the engine, looking for leaks and listening for noise.

If the engine makes excessive noise, turn the ignition OFF immediately. Check the fan assembly, the drivebelts and the exhaust flange bolts. If all these are OK, check the torque converter nuts (automatic transmission). Also, the torque converter sometimes will "scream" if it was not filled with enough fluid prior to reassembly. In this case, run the engine and wait for the pump in the transmission to fill the torque converter. If this doesn't happen fairly quickly, shut the engine down and add some fluid.

Once the engine is running smoothly, allow the engine to idle and adjust the timing. First turn the ignition OFF and loosen the distributor clamp slightly; it should be just loose enough to turn the distributor without allowing the distributor to change position from engine vibration. Attach a stroboscopic timing light to the number one spark plug wire and to the battery terminals on the battery. Start the engine and set the timing. If the vehicle is equipped with points, set the dwell angle. Consult your vehicle's *Haynes Automotive Repair Manual* for exact specifications and additional information.

Turn the ignition OFF and look over the entire engine and compartment for oil leaks, coolant leaks etc. Check the engine oil, coolant, transmission oil and power steering pump oil fluid levels. Add any needed fluids.

On engines equipped with solid lifters, adjust the valves once the engine is warmed up.

Breaking in the engine

Be aware that the rebuilt engine requires a break-in period to allow the new engine parts (rings, pistons, bearings etc.) to seat and wear into their correct position. Go ahead take your wheels for a ride...enjoy your hard work, but don't push the vehicle to any extraordinary limits. It is not going to damage your car to drive normal highway speed on the first few trips. Avoid hard acceleration and do not vary the speed constantly. Listen carefully for any odd sounds or vibrations coming from the rebuilt engine. When you return home, check all the fluid levels and the overall condition of the engine. It's a good idea to change the oil soon after the engine has been put back into action (usually in about 200 miles), since engine break-in puts a lot of metal in the oil. Change the oil again in about 3000 miles, and, if there's not any excessive oil consumption, consider the engine broken in.

What if your new engine uses oil?

Excessive oil consumption after an overhaul can come from a variety of sources, such as leaks and improperly installed rings and valve seals. Also, rings that did not seat properly will cause this problem. Piston rings must create a wear pattern between themselves and the cylinder wall to seal oil out of the combustion chamber. Usually, this happens during the first few hundred miles of the break-in period, so expect some oil consumption during this time. Keep in mind that chrome rings often take longer to seat than cast-iron and moly rings - if you've installed chrome

rings, expect high oil consumption during the first few thousand miles.

If the oil consumption persists for more than a few thousand miles, you may have a ring-seating problem. If, when honing the cylinders, you had a good cross-hatch pattern with the proper-grit stone for your type of piston ring (see Chapter 6), you can usually start pointing your diagnostic finger somewhere else (see Chapter 3). However, if you suspect the cylinder honing may not have been sufficient for the rings you chose, this may be the point of blame. The only truly correct way to deal with this problem is to tear the engine down again and set it up right; however, some shops have devised methods that sometimes work without disassembly. Check around; it may save you a lot of work.

Keep your new engine running for a long time

Now that you have so much time, effort and money tied up in your engine, you'll want to make sure you get the best out of it:

Some people still aren't aware that you can sometimes double or triple the life of an engine just by changing the oil regularly! We normally recommend changing the oil every 3000 miles. Changing it more often won't do any harm; however, we've found this to be a point of diminishing returns. Not changing the oil at regular intervals is asking for trouble! Some people think oil is expensive. You now know how expensive engine parts are.

Checking your air and PCV filters regularly will prevent airflow restriction, increasing fuel economy and making your engine run better. Also, you'll help keep damaging dirt out of the engine.

Make sure you keep your engine tuned up - don't wait until it's running poorly, since this usually puts added strain on the internal engine components, cutting the life of the engine.

We've probably got a *Haynes Automotive Repair Manual* for your particular vehicle that will help you maintain your new engine in top form. Follow the maintenance schedule and your new engine should be churning away for years to come.

8 Related repairs

Since the point of rebuilding the engine is to provide you with a dependable vehicle, and since the components are off the engine anyhow, overhaul time provides an excellent opportunity for checking and replacing or overhauling the engine subsystems so they won't let you down now that you have a reliable engine. These include the carburetor, distributor, fuel pump, starter and alternator.

Carburetor overhaul

Warning: *Gasoline is extremely flammable, so take extra precautions when you work on any part of the fuel system. Don't smoke or allow open flames or bare light bulbs near the work area, and don't work in a garage where a natural gas-type appliance (such as a water heater or clothes dryer) with a pilot light is present. If you spill any fuel on your skin, rinse it off immediately with soap and water. When you perform any kind of work on the fuel tank, wear safety glasses and have a Class B type fire extinguisher on hand.*

1   When it comes time to overhaul the carburetor, you have several options. If you're going to attempt to overhaul the carburetor yourself, first obtain a good quality carburetor rebuild kit (which will include all necessary gaskets, internal parts, instructions and a parts list). You'll also need some special solvent and a means of blowing out the internal passages of the carburetor with air.

2   An alternative is to obtain a new or rebuilt carburetor. They are readily available from dealers and auto parts stores. Make absolutely sure the exchange carburetor is identical to the original. The carburetor model number is stamped on the float bowl or on a metal tag **(see illustration)**. It will help determine the exact type of carburetor you have. When obtaining a rebuilt carburetor or a rebuild kit, make sure the kit or carburetor matches your application exactly. Seemingly insignificant differences can make a large difference in engine performance.

3   If you choose to overhaul your own carburetor, allow enough time to disassemble it carefully, soak the necessary parts in the cleaning solvent (usually for at least one-half day or according to the instructions listed on the carburetor cleaner) and reassemble it, which will usually take much longer than disassembly. When disassembling the carburetor, match each part with the illustration in the carburetor kit and lay the parts out in order on a clean work surface. Overhauls by inexperienced mechanics can result in an

ROCHESTER-TWO JET 2G, 2GC, 2GV, 2GE   2BBL

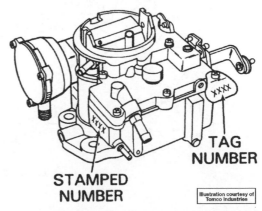

TAG NUMBER

STAMPED NUMBER

Illustration courtesy of Tomco Industries

8.1 The model number is stamped on the float bowl or on a metal tag attached to the carburetor - this number must be used when ordering replacement parts or a new (or rebuilt) carburetor

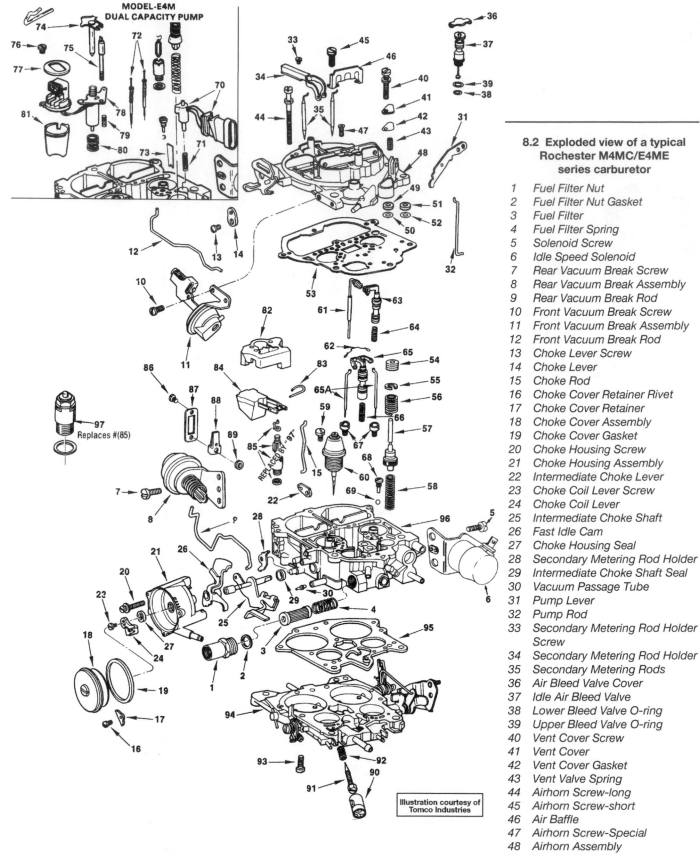

MODEL-E4M
DUAL CAPACITY PUMP

Illustration courtesy of
Tomco Industries

8.2 Exploded view of a typical Rochester M4MC/E4ME series carburetor

1   Fuel Filter Nut
2   Fuel Filter Nut Gasket
3   Fuel Filter
4   Fuel Filter Spring
5   Solenoid Screw
6   Idle Speed Solenoid
7   Rear Vacuum Break Screw
8   Rear Vacuum Break Assembly
9   Rear Vacuum Break Rod
10 Front Vacuum Break Screw
11 Front Vacuum Break Assembly
12 Front Vacuum Break Rod
13 Choke Lever Screw
14 Choke Lever
15 Choke Rod
16 Choke Cover Retainer Rivet
17 Choke Cover Retainer
18 Choke Cover Assembly
19 Choke Cover Gasket
20 Choke Housing Screw
21 Choke Housing Assembly
22 Intermediate Choke Lever
23 Choke Coil Lever Screw
24 Choke Coil Lever
25 Intermediate Choke Shaft
26 Fast Idle Cam
27 Choke Housing Seal
28 Secondary Metering Rod Holder
29 Intermediate Choke Shaft Seal
30 Vacuum Passage Tube
31 Pump Lever
32 Pump Rod
33 Secondary Metering Rod Holder Screw
34 Secondary Metering Rod Holder
35 Secondary Metering Rods
36 Air Bleed Valve Cover
37 Idle Air Bleed Valve
38 Lower Bleed Valve O-ring
39 Upper Bleed Valve O-ring
40 Vent Cover Screw
41 Vent Cover
42 Vent Cover Gasket
43 Vent Valve Spring
44 Airhorn Screw-long
45 Airhorn Screw-short
46 Air Baffle
47 Airhorn Screw-Special
48 Airhorn Assembly

engine which runs poorly or not at all. To avoid this, use care and patience when disassembling the carburetor so you can reassemble it correctly.

4   Because carburetor designs are constantly modified by the manufacturer in order to meet increasingly more stringent emissions regulations, it isn't feasible to include a step-by-step overhaul of each type. You'll receive a detailed, well illustrated set of instructions with any carburetor overhaul kit; they will apply in a more specific manner to the carburetor on your vehicle. An exploded view of a typical four-barrel carburetor is included here **(see illustration)**.

Distributor overhaul

Note: *As an alternative to overhauling a distributor, complete rebuilt distributors are commonly available through well-stocked auto parts stores. Prices are typically very reasonable. If you elect to overhaul the distributor, make sure the necessary parts are available (read through the entire procedure first).*

1   Mount the distributor in a vise, using blocks of wood to protect the housing.

2   Refer to the accompanying exploded views of a point-type distributor **(see illustration)** and breakerless distribu-

49 Pump Stem Seal
50 Pump Stem Seal Retainer
51 T.P.S. Plunger Seal
52 T.P.S. plunger seal Retainer
53 Airhorn Gasket
54 Pump Stem Spacer
55 Pump Spring Retainer
56 Pump Spring
57 Pump Stem Assembly
58 Pump Return Spring
59 Aneroid Assembly Screw
60 Aneroid Assembly
61 Auxiliary Metering Rod
62 Metering Rod Spring
63 Auxiliary Power Piston
64 Auxiliary Power Piston Spring
65 Power Piston Assembly
66 Main Metering Rods
67 Main Jets
68 Pump Discharge Ball Screw
69 Pump Discharge Ball
70 Throttle Position Sensor Assembly
71 T.P.S. Spring
72 Main Metering Rod & Spring
73 Pump Well Baffle
74 Solenoid Plunger
75 Solenoid Lean Mixture Adjusting Screw
76 ECM Connector Screw
77 ECM Connector Gasket
78 ECM Connector & Solenoid Assembly
79 Lean Mixture Screw Spring
80 Solenoid Spring
81 Well Insert
82 Fuel, Bowl Inset
83 Float Hinge Pin
84 Float & Lever Assembly
85 Needle & Seat Assembly
86 Hot Idle Compensator Cover Screw
87 Hot Idle Compensator Cover
88 Hot Idle Compensator Assembly
89 Hot Idle Compensator Gasket
90 Idle Limiter Cap
91 Idle Needle
92 Idle Needle Spring
93 Throttle Body Screw
94 Throttle Body
95 Throttle Body Gasket
96 Main Body
97 Rotary Inlet Valve

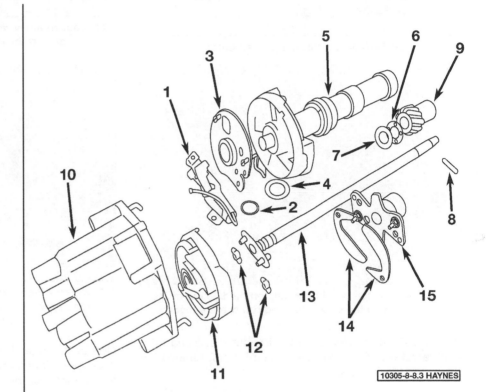

10305-8-8.3 HAYNES

8.3 A typical point-type distributor - exploded view

1   Contact point assembly
2   Retaining ring
3   Breaker plate
4   Felt washer
5   Distributor housing
6   Washer and shim
7   Plastic washer
8   Drive gear roll pin
9   Distributor drive gear
10 Distributor cap
11 Rotor
12 Advance weight springs
13 Mainshaft
14 Advance weights
15 Cam and advance weight base

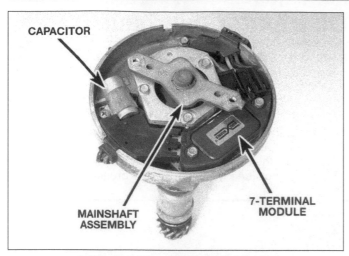

CAPACITOR

MAINSHAFT ASSEMBLY

7-TERMINAL MODULE

8.4 Typical late model HEI distributor

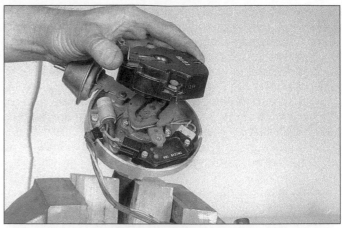

8.5 Loosen the screws (they don't have to be fully removed) and lift the rotor off (breakerless distributor)

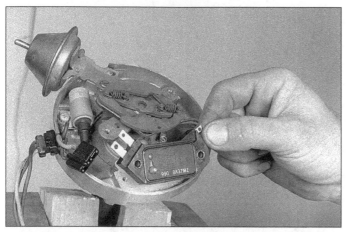

8.6 Remove the screws, lift the ignition module out of the way, then unplug the connector and remove the module

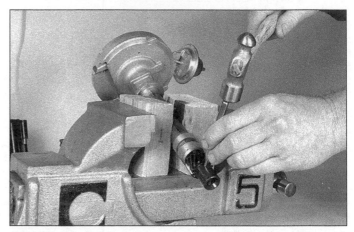

8.7 With the distributor mounted securely in a vise with wood blocks to prevent damage to the housing, drive out the roll pin with the proper size drift

tor **(see illustration)**. Remove the distributor cap and rotor **(see illustration)**. On breakerless distributors, remove the cap with the coil in place.

Point-type distributor

3   Remove the advance weight and springs and (if equipped) the radio interference shield.

Breakerless distributor

4   Remove the ignition module and advance weights **(see illustration)**.

All distributors

5   Mark the distributor gear and shaft so the gear can be reinstalled in the same relationship.
6   Using a drift and hammer, drive out the roll pin and remove the gear and any shims from the shaft **(see illustration)**.
7   Withdraw the shaft from the housing. If there are burrs, corrosion or other deposits on the bottom of the shaft, be sure to sand them off before removing the shaft. Inspect the advance weight assembly for smooth operation, corro-

sion or damage. The shaft and advance weight assembly must be replaced as a unit on later models.
8   Inspect the bushing inside the distributor housing for wear and signs of excessive heat concentration (bluing) and the housing itself for cracks and wear. Replace the complete distributor housing if the bushings or housing are damaged.

Point-type distributor

9   Remove the cam weight base assembly.
10   Remove the screws and lift off the vacuum advance unit
11   Remove the snap-ring and remove the breaker plate assembly. Remove the contact points and condenser, followed by the felt washer and plastic seal under the breaker plate.
12   Inspect the parts for wear, distortion and cracks, replacing as necessary.
13   Assembly is the reverse of disassembly, but make sure to first fill the lubricating cavity in the housing with multipurpose grease and to grease the advance mechanism cam weight base. Put a few drops of engine oil on the lubri-

8.8 Check the starter motor pinion for freedom of movement by rotating it in both directions with your fingers - it should rotate freely in one direction

8.9 Check the armature for freedom of rotation by prying the pinion with a screwdriver - the armature should rotate freely

cating wick before installing it. Use a new plastic seal and felt washer and a new roll-pin when installing the gear.

Breakerless distributor

14 Remove the "C" clip and lift off the pickup coil assembly.
15 Remove the screws and detach the vacuum advance unit.
16 Remove the capacitor and the wiring harness from the housing.
17 Inspect the parts for wear, distortion and cracks, replacing as necessary.
18 Assembly is the reverse of disassembly, but make sure to first fill the lubricating cavity in the housing with multipurpose grease if it's empty or low on lubricant and to grease the advance mechanism. Lubricate the electrical connectors with petroleum jelly and coat the bottom of the ignition module with silicone grease before installation.

Fuel pump overhaul

The fuel pump on some earlier models can be disassembled and rebuilt, while on later models the pump is sealed and must be replaced as a unit. Rebuildable pumps have screws around the pump body which can be removed to allow disassembly. Prior to beginning work, obtain a rebuild kit from your dealer. As you disassemble the fuel pump, lay the parts out in order on a clean work surface. This way the pump components can be reassembled in the proper sequence. **Warning:** *Gasoline is extremely flammable, so take extra precautions when you work on any part of the fuel system. Don't smoke or allow open flames or bare light bulbs near the work area, and don't work in a garage where a natural gas-type appliance (such as a water heater or clothes dryer) with a pilot light is present. If you spill any fuel on your skin, rinse it off immediately with soap and water. When you perform any kind of work on the fuel tank, wear safety glasses and have a Class B type fire extinguisher on hand.*

Starter overhaul

1 Wipe the starter clean with a shop rag and solvent and place it on a work bench.
2 Check the pinion for damage and rotate it **(see illustration)**. It should only rotate easily in one direction. If it rotates easily in both directions, and the armature does not move, the overrunning clutch is faulty.
3 Check the armature for freedom of rotation by rotating the pinion with a screwdriver in the overrunning clutch's locked position **(see illustration)**. Tight bearings or a bent armature shaft will prevent the armature from turning easily. If the armature does not turn freely, disassemble the motor immediately. If the armature does turn freely, perform a no-load test before you disassemble the motor.
4 To no-load test the starter, simply hook up a 12-volt battery, using jumper cables, as shown and jump the solenoid terminals with a screwdriver **(see illustration)**. Don't run the starter for more than a few seconds at a time

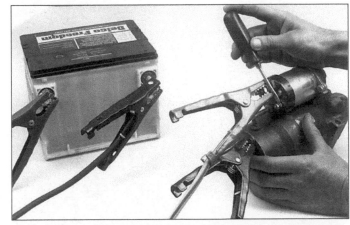

8.10 To do a no-load bench test of the starter, hook up a 12-volt battery - positive to the battery terminal, negative on the starter case - connect a jumper wire or an old screwdriver (as shown here) between the battery terminal and the solenoid control circuit terminal

8.11 Disconnect the field strap from the terminal on the solenoid

8.12 Remove the bolts which attach the solenoid to the starter drive housing

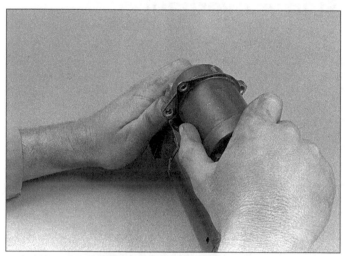

8.13 Twist the solenoid clockwise . . .

8.14 . . . and separate it from the starter housing - remove the return spring and set it aside

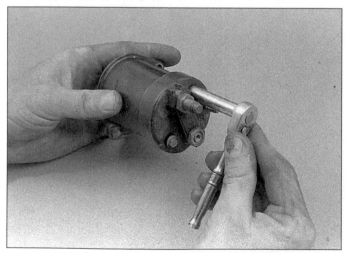

8.15 Remove the two solenoid cover screws

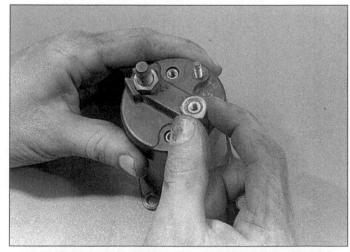

8.16 Remove the nuts from the motor terminal (usually marked "B") and the switch terminal - terminal studs have welded leads, so avoid twisting the terminal when loosening the nut

8.17 Remove the solenoid cover and inspect the contacts - if they're dirty, clean them with contact cleaner; if they're burned or damaged, replace the solenoid

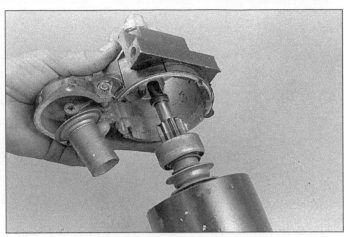

8.18 Remove the two through-bolts from the starter and remove the drive end housing

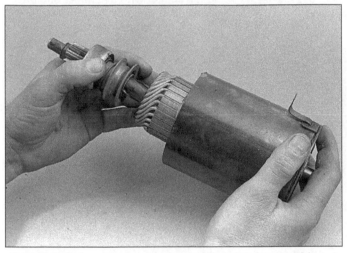

8.19 Pull the armature and overrunning clutch assembly out of the field frame

8.20 Remove the commutator end frame

and make sure the battery is in good condition. **Warning:** *Keep your fingers away from the rotating pinion gear!*

Disassembly

5 Follow the procedure shown in the **accompanying photos** to disassemble the starter.

6 Clean the overrunning clutch with a clean cloth. Clean the armature and field coils with electrical contact cleaner and a brush. Do not clean the clutch, armature or field coils in a solvent tank or with grease cutting solvents - they'll dissolve the lubricants in the clutch and damage the insulation in the armature and field coils. If the commutator is dirty, clean it with 00 sandpaper. Never use emery cloth to clean the commutator.

7 Test the operation of the overrunning clutch. The pinion should turn freely in the overrunning direction only. Check the pinion teeth for chips, cracks and excessive wear. Replace the clutch assembly if damage or wear is evident. Badly chipped pinion teeth may indicate chipped teeth on the flywheel ring gear. If this is the case, be sure to check

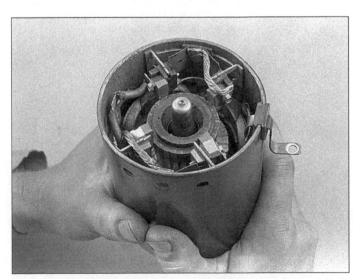

8.21 This is what the brushes, brush holders and springs look like when assembled properly

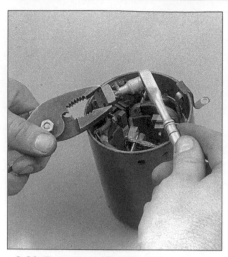

8.22 To remove the brushes, remove the small screw attaching the lead to each brush

8.23 Inspect the brushes for wear - replace the set if any are worn to one-half their original length. Judging from this brush, the brushes in this starter should be replaced

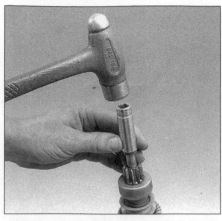

8.24 To remove the overrunning clutch from the armature shaft, slide an old socket onto the shaft so that it butts against the edge of the retainer, then tap the socket with a hammer and drive the retainer towards the armature and off the snap-ring

8.25 Remove the snap-ring from its groove in the shaft with a pair of pliers or screwdriver - if the snap-ring is badly distorted, use a new one during reassembly

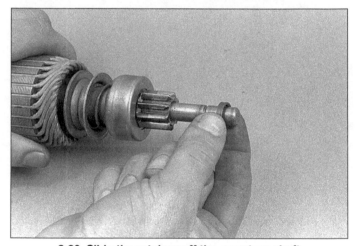

8.26 Slide the retainer off the armature shaft

8.27 Slide the drive assembly off the shaft - reassembly of the starter is the reverse of disassembly

the ring gear and, if necessary, replace it.

8   Inspect brushes for wear. Replace them as a set if any are worn to one-half their original length. The brush holders must hold the brushes against the commutator. Make sure they're not bent or deformed.

Armature

9   Check the fit of the armature shaft in the bushings at each end of the motor housing. The shaft should fit snugly in the bushings. If the bushings are worn, replace them or obtain a rebuilt starter motor.

Commutator

10   Check the commutator for wear. If it's heavily worn, out-of-round or the insulation strips between each segment are high, the armature should be turned on a lathe and the insulation strips should be undercut. This operation is best performed by a properly equipped automotive machine shop that's familiar with the procedure. It may be quicker

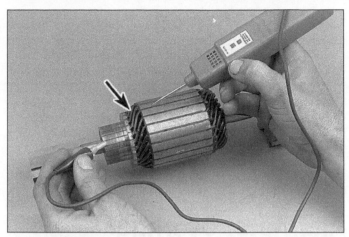

8.28 Checking for a short between a commutator segment and the armature - note how each commutator segment and associated winding are joined by a conductor (arrow)

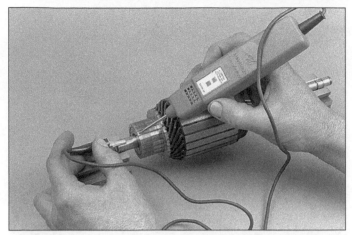

8.29 Checking the armature and commutator for a ground

and easier to obtain a rebuilt starter.

11 Using an ohmmeter or a continuity tester, check for shorts between each commutator segment and the armature **(see illustration)**. Replace the armature if there are any shorts. If there's no continuity, there's and open in the winding. The most likely places for opens to occur are at the points where the armature conductors join the commutator segments. Inspect these points for loose connections. Poor connections cause arcing and burning of the commutator segments as the starter motor is used. If the segments aren't too badly burned, the damage can often be repaired by resoldering the conductors to the segments and cleaning up the burned material on the commutator with 00 sandpaper.

12 Using an ohmmeter or a continuity tester, check for continuity between each commutator segment and the segment to its immediate right or left. There should be continuity.

13 Shorts are sometimes produced by carbon or copper dust (from the brushes) between the segments. They can usually be eliminated by cleaning out the slots between the segments. If a short persists, you'll need to take the armature to a properly equipped shop to be tested on a growler. Often, it's easier to obtain a rebuilt starter.

14 Check for grounds. These often occur as a result of insulation failure brought about by overheating the starter motor by operating it for extended periods. They can also be caused by an accumulation of brush dust between the commutator segments and the steel commutator ring. Grounds in the armature and commutator can be detected with an ohmmeter or a continuity checker. Touch one probe of the test instrument to the armature shaft and touch the other probe to each of the commutator segments **(see illustration)**. If the instrument indicates continuity at any of the segments, the armature is grounded. If cleaning won't correct the ground problem, replace the armature (often it's easier to simply obtain a rebuilt starter).

Field coils

15 Using an ohmmeter or a continuity tester, place one probe on each field coil connector **(see illustration)**. The instrument should indicate continuity. If there's no continuity, there's an open in the field coils. It's best to obtain a rebuilt starter since field coil replacement normally requires special tools.

16 Place one probe of an ohmmeter or continuity tester on one of the field coil connectors **(see illustration)**. Place the

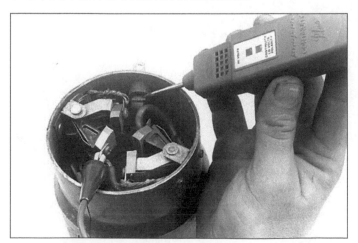

8.30 Checking for an open in the field coils

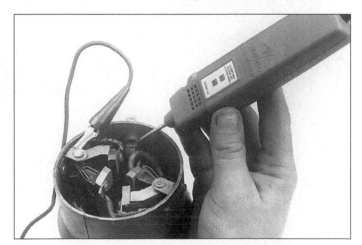

8.31 Checking for a grounded field coil

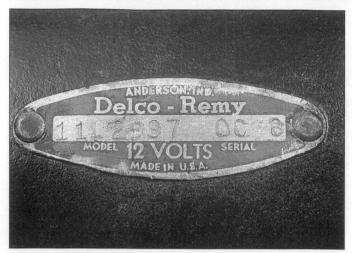

8.32 A typical generator identification tag

Generator overhaul

Many older (pre-1970's) vehicles use generators. Modern vehicles are equipped with alternators to handle the higher current demands of today's electrical systems. If the generator is to be replaced or overhauled, refer to identification tag riveted to the housing to make sure the replacement unit is the same or that the correct parts are obtained **(see illustration)**. A generator is less complicated than an alternator, with fewer parts. Consequently, overhaul is a simpler procedure, consisting of disassembly, replacement of worn parts and reassembly **(see illustration)**.

Alternator overhaul

other probe on the starter frame. Disconnect the shunt coil ground, if applicable, before you do this check. If the instrument indicates continuity, the field coils are grounded; obtain a rebuilt starter.

Note 1: Special tools are needed to overhaul an alternator. These tools include an ohmmeter, soldering gun and special pullers and installers. Before disassembling the alternator, check the availability and prices of overhaul kits and also the prices of new and rebuilt alternators. Often you'll find it easier - and sometimes less expensive - to obtain a rebuilt unit.

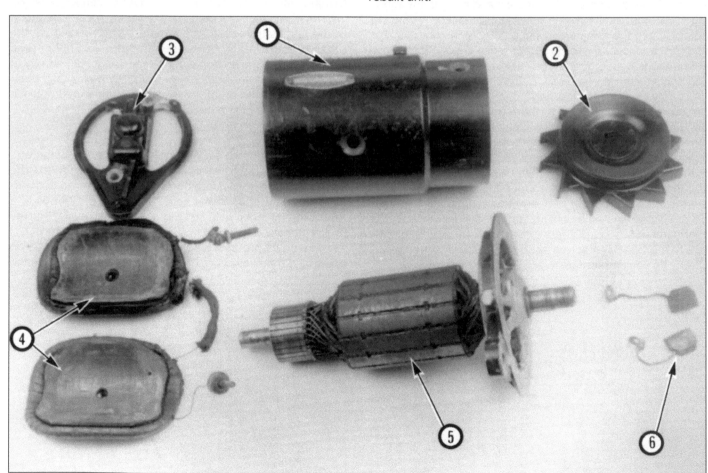

8.33 A typical generator - disassembled

1	Field frame	3	Commutator end frame	5	Armature
2	Fan pulley	4	Field windings	6	Brushes

8.34 Some late model alternators cannot be rebuilt - they can be identified by special bolts (arrow) or rivets

8.35 Paint or scribe a mark across the length of the alternator housing to ensure correct reassembly

8.36 On most models, you can remove the fan pulley with an Allen and box-end wrench

8.37 After the fan pulley has been removed, some alternators will have a spacer

8.38 Remove the through-bolts (arrows)

Note 2: *Some later model alternators are not intended to be disassembled and rebuilt. These usually have through-bolts with special heads or rivets in place of through-bolts* **(see illustration)**

Note 3: *The following procedure is general in nature. If the instructions you receive with your overhaul kit differ from the procedure that follows, assume the instructions are correct.*

Disassembly

1   Scribe or paint a line across the alternator housing to ensure correct reassembly **(see illustration).**
2   Remove the pulley. On most models the pulley can be removed with common hand tools **(see illustration).** Most alternators have a spacer below the pulley **(see illustration).**
3   Remove the housing through-bolts **(see illustration).**
4   Carefully separate the end frames **(see illustration)**

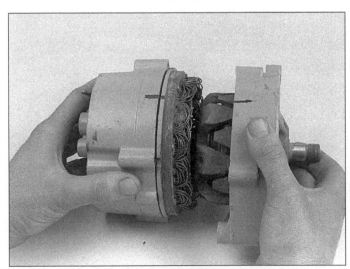

8.39 Carefully separate the end frames

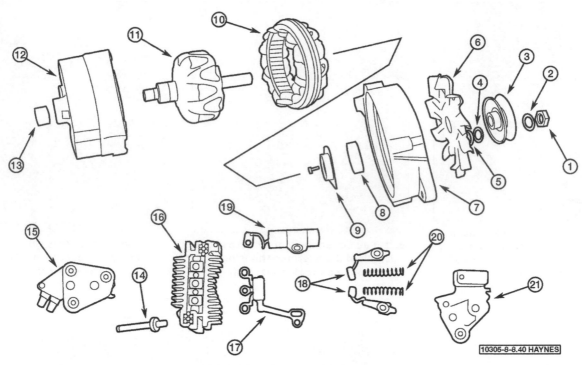

8.40 A typical early model alternator

1	Nut	8	Bearing	15	Voltage regulator
2	Washer	9	Plate	16	Rectifier bridge
3	Pulley	10	Stator	17	Diode Trio
4	Washer	11	Rotor	18	Brushes
5	Collar	12	Slip-ring end frame	19	Capacitor
6	Fan	13	Bearing	20	Brush springs
7	Drive end frame	14	Terminal component stud	21	Brush holder

and refer to the accompanying exploded view **(see illustration)**.

5   Remove the brush holder and/or regulator **(see illustration)**.

6   Special pullers and installers are recommended to remove and install the bearings. However, an automotive machine shop can usually do this for a reasonable charge.

Inspection

7   After the alternator has been completely disassembled, clean all parts with electrical contact cleaner. Do not use de-greasing solvents; they can damage electrical parts. Look for cracks in the case, burned spots, worn brushes **(see illustration)**, worn slip rings **(see illustration)** and

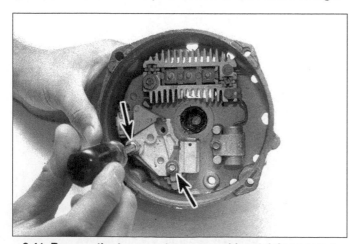

8.41 Remove the two regulator assembly retaining screws

8.42 A typical example of worn out brushes

8.43 A typical example of worn out slip rings

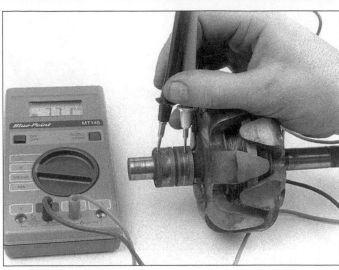

8.44 Check for continuity between the two slip rings - if there's no continuity, the rotor is open and should be replaced

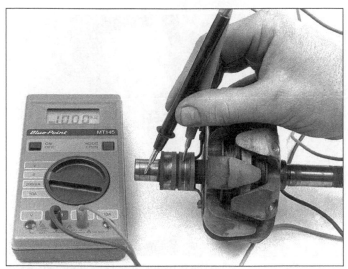

8.45 Check for continuity between each slip ring and the rotor shaft - if there's continuity, the rotor is grounded and should be replaced

8.46a Check for continuity between these two stator terminals . . .

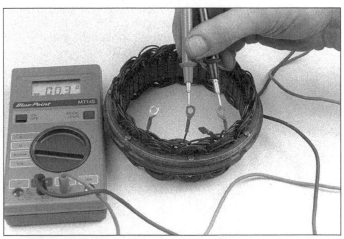

8.46b . . . and these two - if there's continuity, there is an open in the windings and the stator should be replaced

other parts exhibiting wear or other damage. Replace parts as necessary.

Testing

8   With an ohmmeter, check the rotor for opens **(see illustration)** and grounds **(see illustration)**. Replace the rotor or obtain a rebuilt alternator if either condition is present.

9   If the rotor checks out OK, clean the slip ring with 400 grit or finer polishing cloth.

10   Check the stator for opens and grounds **(see illustrations)**. Replace the stator or obtain a rebuilt alternator if either condition is present.

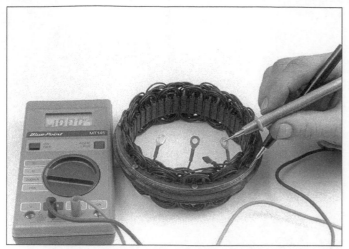

8.46c Check for continuity between each stator terminal and ground - if there's no continuity, the stator is grounded and should be replaced

8.47a If the alternator has a diode trio like this, connect an ohmmeter as shown, then reverse the ohmmeter leads - one high and one low reading indicates a good diode trio - if the readings are about the same, the diode trio should be replaced

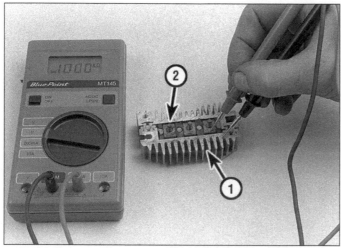

8.47b If the alternator has a reflector bridge like this, connect an ohmmeter as shown - one lead to the insulated heat sink (1) and the other to each of the metal clips; take a reading then reverse the ohmmeter leads and take another reading - you should get one high and one low reading at each clip - if the readings are about the same replace the rectifier bridge. Next, repeat this test between the three clips and the grounded heat sink (2)

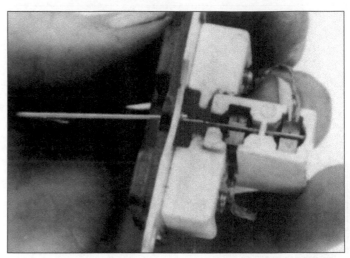

8.47c Before installing the brushes, insert a paper clip, as shown, to hold the brushes in place during installation - after installation simply pull the paper clip out

11 Check each diode or diode trio and rectifier bridge, as described in the **accompanying illustrations**. Replace any faulty parts.

Assembly

13 Assembly is the reverse of disassembly.

14 Before installing the brush holder, insert a paper clip to hold the brushes in place during assembly **(see illustration)**. Remove the clip after assembling the alternator.

9 Improving performance and economy

Introduction

In these times of ever-increasing fuel costs, the beleaguered enthusiast is constantly reminded of the trade-offs required to drive a performance vehicle. It takes fuel to produce horsepower, and the faster you go, the more you need.

But performance and economy don't have to be mutually exclusive. By properly matching components and careful tuning, you can improve the power and efficiency of the drivetrain to obtain the best of both worlds.

Automotive designs are, by necessity, fraught with compromises. Factory engineers must allow for wide variations in production line tolerances, driving techniques, low octane fuel, carbon build-up, wear, emissions certification, neglect and lack of maintenance while keeping costs down.

Stock production passenger cars and light trucks are designed for a balance between everyday stop-and-go driving around town and cruising on the highway. The engines and drivelines are optimized for low and mid-range power rather than high rpm peak horsepower.

Engines are basically air pumps that mix fuel and air and produce power from the combustion. Anything you can do to increase the flow of air (assuming the fuel system is capable of delivering sufficient fuel in the correct proportions) through the engine will increase power. Other ways to increase power and/or economy are to reduce weight, friction and drag.

Every engine is designed to operate most efficiently in a certain speed range (measured in revolutions per minute [rpm]) . The length and diameter of the intake and exhaust ports and the intake and exhaust manifolds (or headers) help determine the power band of the engine. Long, small-diameter intake and exhaust runners improve low-rpm torque and decrease high-rpm power. Conversely, short, large cross-section passages favor high rpm power.

The type and rating of intake and exhaust systems, the camshaft, valve springs and lifters, ignition, cylinder heads, valve diameters and bore/stroke relationship are matched at the factory to ensure a good combination of economy,

power, driveability and low emissions. Additionally, the transmission characteristics, differential gear ratio and tire diameter must all work in harmony with the engine.

For street driving, low and mid-range torque is much more useful (and economical) than theoretical ultimate horsepower at extremely high rpm. A street-driven engine that produces high torque over a wide range of rpm will deliver more average horsepower during acceleration through the gears than an engine that delivers higher peak power in a narrow range of rpm.

Heavy vehicles with relatively small engines should have lower gearing (higher numerically) than light vehicles with relatively large engines. Also, the engine in a heavy vehicle should be optimized for maximum torque in the low and middle rpm range, since it takes more torque to get a heavy vehicle moving and to accelerate it.

New cars and trucks have low numerical axle ratios, lock-up torque converters, overdrives and more ratios in the transmissions to get good mileage and acceleration. One of the best ways to improve performance and economy at the same time on an older vehicle is to install a transmission with more forward gears and a differential with better ratios than stock. Frequently, these parts are available from late-model vehicles at wrecking yards for reasonable prices.

Most racing engines run in a narrow range at high rpm and don't need the flexibility low-speed torque provides. Many hot rodders succumb to the temptation of putting a radical racing camshaft or a huge carburetor on an otherwise stock engine. This increases the theoretical air flow capacity in one part of the engine without changing the flow characteristics of other components. Since the components aren't matched, intake air charge velocity slows down and fuel doesn't mix properly with the air. The engine no longer has an optimum rpm band. This results in a gas-guzzling slug that isn't as fast as stock.

Torque, usually measured in foot-pounds (ft-lbs) in the USA, is a measure of the twisting force produced by the engine. Horsepower is a measure of work done by the engine. Torque (in ft-lbs) multiplied by engine speed (rpm) divided by 5,250 equals horsepower.

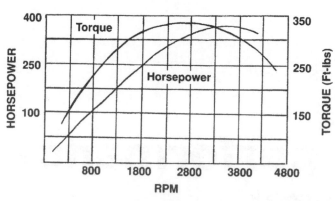

9.1 Note how torque drops off before horsepower on this typical engine

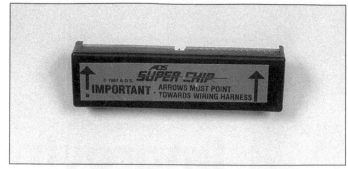

9.2 Special chips are available for computer equipped vehicles

Engines produce the most power from a given amount of fuel at their peak torque. This is the rpm the factory engine design is optimized for. Peak horsepower is achieved by spinning the engine faster than this most- efficient speed. The torque peak always occurs at a lower rpm than the peak horsepower **(see illustration)**. Horsepower peaks when the gains made by running faster are balanced by the losses caused by running above the optimum speed the engine components are tuned for.

You can tell a lot about an engine from its power specifications. On a high-performance engine, maximum horsepower will usually be higher than maximum torque, and peak power will occur at a relatively high rpm. Also, as a general rule, high-performance street engines put out approximately one horsepower per cubic inch or better. For example, a hypothetical standard engine might have 300 cubic inches of displacement, maximum torque of 275 ft-lbs @ 3,000 rpm and 200 hp @ 4,200 rpm. The high-performance version with the same displacement might have 245 ft-lbs @ 3,800 rpm and 325 hp @ 5,600 rpm. **Note:** *Late-model (about 1972 and later) engine horsepower ratings are net (with accessories connected) whereas earlier models are rated for gross or brake horsepower (without accessories). Net ratings tend to be lower, but more realistic than the gross ratings.*

Before you select components to modify your vehicle, you should plan out realistically what you want it to do. Your engine must be in excellent condition to start with, otherwise it will quickly self-destruct. Check the condition of the engine thoroughly, following the procedures described in Chapter 3. If necessary, rebuild it now; you can include modifications during the overhaul and it will cost less than if you did the work separately.

Find out what the factory rated horsepower and torque are, and at what rpm the peaks occur. Then determine what rpm the engine is turning at your usual highway cruising speed and what the gearing is. If your vehicle isn't equipped with a tachometer, temporarily connect a test meter with long wires run into the interior. To determine the axle ratio, check the tag on the differential.

Once you know these things, you can decide which

way to go. Generally, if you modify the vehicle so it has to run at higher rpms and/or you want a large increase in power, expect to sacrifice a considerable amount of fuel economy and reliability.

Some of the more popular ways to dramatically raise power output are supercharging, turbocharging, nitrous oxide injection or by swapping a much larger displacement engine into the vehicle. There are many entire books devoted to each of these methods, and these subjects are beyond the scope of this Chapter.

Depending on the year and model of your vehicle, you may be able to make substantial improvements in mileage through careful tuning, by changing axle ratios, tires, intake and exhaust modifications, camshaft replacement and ignition improvements. Vehicles from the 1950s through the 1970s are most responsive to these changes, which must be carefully planned and coordinated.

Newer computer-controlled models have many of these changes incorporated in them already and get better mileage than their predecessors. They are so sensitive to modifications that even a change in tire diameter can affect the driveability of the vehicle.

There are very few modifications that can be done to computer-controlled vehicles without making them violate emission laws. Several aftermarket manufacturers produce intake manifolds, camshafts, exhaust systems and computer chips **(see illustration)** that can increase the performance of late-model vehicles. Shop carefully and read the fine print to determine computer compatibility before you purchase any components.

If you are planning to rebuild your engine, you may want to make many modifications during the overhaul. While the engine is apart, you can easily change the cylinder heads, pistons, connecting rods, crankshaft and camshaft. Modified cylinder heads can provide substantial gains in high rpm power. For a mild street engine, a good three-angle valve job and matching the intake ports to the intake manifold will make it run better without sacrificing driveability. Older engines can benefit from the addition of hardened valve seats and special valves to enable them to run on low-lead or unleaded gasoline. See Chapter 5 for additional information.

High-compression pistons improve power and efficiency at all speeds, but if you exceed about 9:1, premium

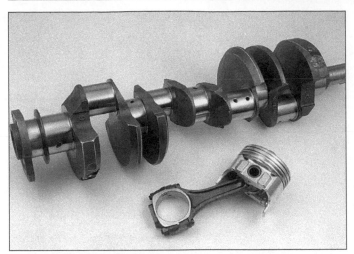

9.3 Long-stroke crankshafts and oversize pistons can boost torque and horsepower by increasing displacement

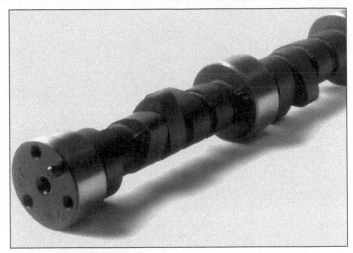

9.4 As the camshaft turns, the lobes push the valve lifters up

fuel is necessary. Flat-top pistons produce a better flame front in the combustion chamber than dished (concave) ones. Forged pistons are stronger than cast pistons; however, cast pistons work fine on the street.

Longer stroke crankshafts with matching connecting rods and large-bore pistons **(see illustration)** can increase horsepower without sacrificing driveability or low-end torque. However, if you intend to build a high rpm engine, don't get carried away with this: long-stroke engines can limit high-rpm potential.

Before you assemble the engine, have an automotive machine shop blueprint and balance the parts; it's a way of finding extra horsepower that doesn't require more fuel.

In this Chapter we will discuss the pros and cons of various component changes and how they affect other parts of the vehicle. Usually, if you change one part, you'll have to modify or replace others that work in conjunction with it. Check the fuel mileage of your vehicle and measure performance carefully with a stop watch before and after each modification to determine its effect. Test before and after under the same conditions on the same roads to ensure accuracy.

Many books and articles have been written on how to turn your street vehicle into a race car by spending a fortune. Of course, that leaves you back at square one, because you don't have a vehicle to drive to work anymore. We will attempt to limit the discussion to modifications that can be done at home for a reasonable cost and that will not prevent the vehicle from being used as daily transportation.

Make sure any modifications are legal according to the latest federal, state and local laws. Many states conduct "smog" inspections for street-driven vehicles. Removing or modifying emission control systems is often illegal, and usually results in little or no improvements in power or economy. Besides, we all have to breathe the air. Many jurisdictions forbid loud exhausts, too. Be sure to check with your state department of motor vehicles and state or local police regarding current regulations.

Camshaft selection

Introduction

The camshaft is the mechanical "brain" of the engine. It determines when and how fast the valves will open and close, and how long they stay open by pushing up valve lifters **(see illustration)** with elliptical (egg-shaped) lobes as it turns **(see illustration)**.

The camshaft, more than any other single part, determines the running characteristics (or "personality") of the engine. A single camshaft design can not provide maximum power from idle to redline. Like everything else in a motor vehicle, camshaft design is a compromise. If a camshaft is designed to produce gobs of low-end torque, good driveability and fuel economy, it must give up some peak horsepower at high rpm. Conversely, camshafts designed for high rpm run poorly at idle and lower engine speeds.

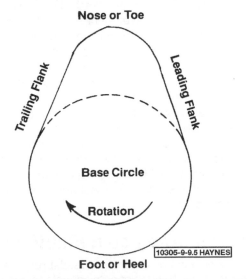

9.5 Each portion of the camshaft profile has a specific name and function

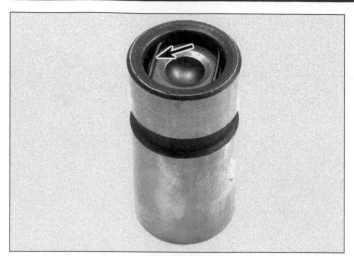

9.6 Hydraulic lifters can be identified by the retainer clips in the top (arrow)

9.7 Aftermarket roller lifters

Valve lifter types

Before you can make an informed choice on camshaft selection, you need to know some basic design parameters and terminology. Camshafts must be designed to work with a certain type of valve lifter and must not be used with any other type. There are three basic types of valve lifters: mechanical, hydraulic and roller. Mechanical lifters, also known as solid or flat tappets, are the oldest, simplest and least expensive type. Because of their light weight, mechanical lifters allow an engine to rev slightly higher before the valves float. The main disadvantages of mechanical lifters are the necessity of frequent valve adjustments and the noise they produce.

Hydraulic lifters **(see illustration)** are the most common type used in V8 engines. They have a small internal chamber where engine oil collects and a check valve to prevent backflow. This feature allows the lifter to automatically compensate for differences in valve lash, or clearance. Standard hydraulic lifters are relatively inexpensive and are maintenance-free; however, at high rpm, they tend to pump up and float the valves. Special high-performance lifters are available which extend the rpm range high enough for virtually any street machine. Hydraulic lifters are the most popular type of lifter used in performance street engines, and work well in most applications.

Roller valve lifters **(see illustration)** are the best and most expensive type available. They increase horsepower and improve fuel economy by reducing friction. Roller lifters are available in both solid and hydraulic versions.

If you can afford it, buy a roller camshaft and lifters. Hydraulics are next best and mechanical are least desirable for a street engine.

Understanding specifications

Every engine has a certain speed it runs best at, which is a result of the "tuning" of components to achieve an optimum flow velocity of air/fuel mixture.

The main reason engines don't run at maximum efficiency throughout the power band is because air has mass, and therefore inertia. As engine speed increases, the amount of time available for the gases to enter and exit the combustion chamber becomes less. Camshaft designers compensate for this by opening the valves earlier and holding them open longer. But the valve timing that works well at low speed is inefficient at high speed.

Several quantifiable factors determine camshaft characteristics. The most important and widely advertised items are LIFT, DURATION and OVERLAP **(see illustration).**

Lift

Lift is simply the amount of movement imparted by the camshaft lobe. Lift specifications can be confusing because the rocker arms multiply the actual lift at the camshaft lobe by a ratio of approximately 1:1.5 to 1:1.7. Most camshaft manufacturers provide "net" lift specifications which is the maximum amount of lift (or movement) that occurs at the valve. Actual lobe lift measured at the camshaft is considerably less than net lift. Up to about 0.5 in. net lift, more lift produces more power. Beyond this point, there are diminishing returns. High lifts also result in greater wear rates and premature failure of valve train components.

Duration

Duration indicates how long a valve stays open and is measured in degrees of crankshaft rotation (remember that camshafts turn at one half of crankshaft speed). Long duration increases high rpm power at the expense of economy, exhaust emissions and low-end power.

Comparing duration of different camshafts is difficult because not all manufacturers use the same method of measuring. Some companies measure duration from the exact point where the valve lifts off the seat. This produces a higher number, but in practice, the fuel/air mixture does not begin to flow significantly until the valve has lifted a certain amount.

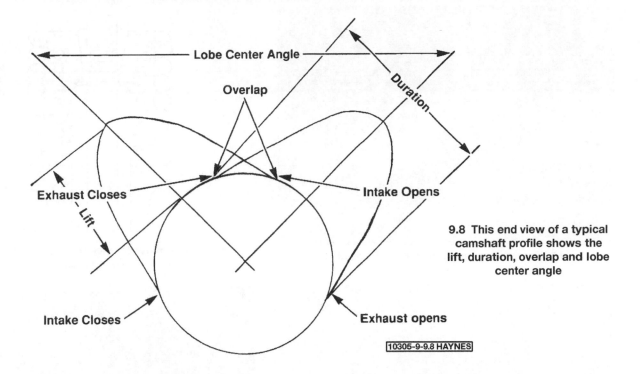

Lobe Center Angle

Overlap

Duration

Exhaust Closes

Lift

Intake Opens

Intake Closes

Exhaust opens

9.8 This end view of a typical camshaft profile shows the lift, duration, overlap and lobe center angle

10305-9-9.8 HAYNES

Most camshaft experts have agreed to measure duration starting and ending at 0.050 in. lift. This method produces a smaller number that is more representative of flow characteristics. For street use, cams having about 230-degrees of duration (measured at 0.050 in. lift) work well. Be sure you know which method of measurement is being used when you compare camshafts from different manufacturers.

Overlap

Overlap is a measurement of crankshaft rotation in degrees during which both the intake and exhaust valves are open. Like duration, long overlap also increases high rpm power at the expense of economy, exhaust emissions and low-end power.

Two factors influence valve overlap specifications. First and easiest to understand is the amount of valve duration. Second is the lobe centerline angle or lobe displacement of the camshaft.

Other factors

The lobe center angle varies indirectly with the overlap. That is, when overlap increases, lobe center angle decreases, and vice versa. Increasing the angle generally increases low-end torque and reducing the angle improves high-rpm horsepower.

Another area of design that affects camshaft characteristics is the lobe profile. The rate of lift, acceleration and rate of valve closing are determined by the shape of the lobes and affect the way the engine runs. By opening and closing the valves more quickly, cam designers can get more flow from a given amount of duration.

Camshafts and valve train components must be matched to work together properly. In addition, you must carefully match the camshaft/valve train to the other components used on the engine and vehicle, especially the intake and exhaust, gearing and transmission.

Most camshaft manufacturers have technical departments that help customers determine the best camshaft and components for their specific application. In order make accurate recommendations, they need complete information on the vehicle. If you have a heavy vehicle with a relatively small engine, you must be conservative with the camshaft and other components. Follow the recommendations of the camshaft manufacturers; they have done a lot of testing and research.

Modified camshaft timing can result in valve-to-piston interference and severe engine damage. Be sure to check the following when you install a camshaft with different specifications than the original one:

Valve spring coil bind

Whenever a camshaft with higher lift than stock is installed, the springs should be checked for coil bind. Due to the increased travel, the valve spring coils may hit together (bind), causing considerable damage.

Perform this check with the new camshaft and lifters in place and the valve covers off. The cylinder heads, rocker arms and push rods must be in place and adjusted. Using a socket and breaker bar on the front crankshaft bolt inside the lower pulley, carefully turn the crankshaft through at least two complete revolutions (720-degrees). When the valve is completely open (spring compressed), try to slip a

9.9 Use a feeler gauge to check for coil bind

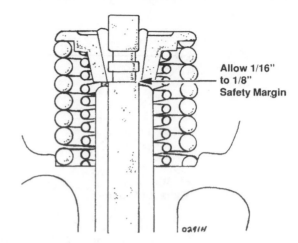

Allow 1/16"
to 1/8"
Safety Margin

9.10 Check for interference between the valve guide and spring retainer

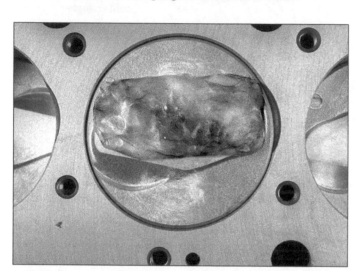

9.11 Press modeling clay onto a piston where the valves come close to the piston crown

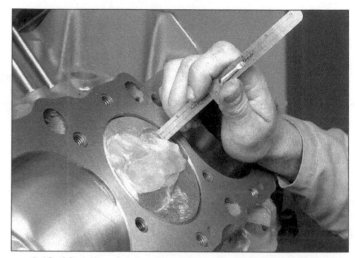

9.12 After the clay is compressed, cut through it at the thinnest point and measure that part

0.010 in. feeler gauge between each of the coils **(see illustration)**. It should slip through at least two or three of the coils. If any spring binds, stop immediately and back up slightly. Then correct the problem before continuing. Usually, the valve springs must be replaced with special ones compatible with the camshaft.

Spring retainer-to-valve guide clearance

Sometimes high-lift camshafts will cause the valve spring retainers to hit the valve guide. To check for this, rotate the crankshaft as described above and check for guide-to-retainer interference **(see illustration)**. There should be at least 1/16 in. clearance.

Piston-to-valve clearance

Remove a cylinder head and press modeling clay onto the top of a piston **(see illustration)**. Temporarily reinstall the cylinder head with the old gasket and tighten the bolts. Install and adjust the rocker arms and pushrods for that cylinder. Rotate the crankshaft through at least two com-

plete revolutions (720-degrees). Remove the cylinder head and slice through the clay at the thinnest point **(see illustration)**. Measure the thickness of the clay at that point. It should be at least 0.080 in. thick on the intake side and 0.10 in. thick on the exhaust side. If clearance is close to the minimum, check each cylinder to be sure a variation in tolerances doesn't cause valve-to-piston contact.

Degreeing the camshaft

Every manufactured part has a design tolerance, and when several parts are put together, these tolerances can combine to create a significant error, called tolerance stacking. A lot of power can be lost by assembling an engine without checking and, if necessary, correcting the camshaft timing.

Installation tips

The first step is to find true Top Dead Center (TDC); original factory marks can be off by several degrees. Remove the number one spark plug. Using a socket and

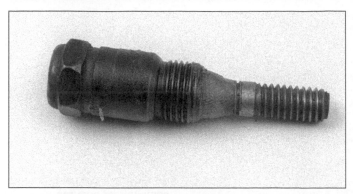

9.13 You can buy a positive stop from a speed shop or fabricate one yourself

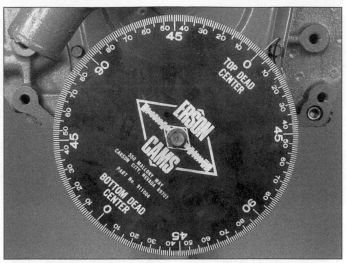

9.14 Mount a degree wheel on the front of the crankshaft and install a pointer as shown

9.15 Mount a dial indicator in line with the pushrod, as shown

breaker bar, rotate the crankshaft slowly in a clockwise direction until air starts to blow out of the spark plug hole. Stop rotating the crankshaft at this point.

Obtain a positive stop tool **(see illustration)**. Screw it into the number one cylinder spark plug hole; it prevents the piston from going all the way to the top of the bore during this procedure.

Mount a degree wheel securely to the front of the crankshaft and fabricate a pointer out of coat hanger wire **(see illustration)**. Rotate the crank shaft very slowly in the same direction (clockwise when viewed from the front) until the piston touches the stop. **Note:** *Use a large screwdriver on the teeth of the flywheel/driveplate to turn the crankshaft; otherwise you may disturb the degree wheel. Note what number the pointer aligns with and mark the degree wheel with a pencil.*

Now rotate the crankshaft slowly in the opposite direction until it touches the stop. Again note what number the pointer aligns with and mark the degree wheel with a pencil. True TDC is exactly between the two marks you made.

Double-check your work, remove the piston stop, then rotate the crankshaft to true TDC and check the factory mark. If it is off by a degree or more, correct the factory timing marks.

Now that you've found true TDC, you can check camshaft timing in relation to your crankshaft. Advancing

the camshaft in relation to the crankshaft improves low-end performance and retarding the camshaft benefits high-rpm performance at the expense of low-end performance. Whenever cam timing is changed, ignition timing must be reset, and valve-to-piston clearance should be checked if a large change is made.

Read the specification tag included with the camshaft, it should have the opening and closing points (in degrees) of the intake and exhaust valves at 0.050 in. lift. **Note:** *This is the industry standard; if the cam manufacturer uses a different lift, follow their recommendations.*

Begin this check with the engine on true TDC and the degree wheel set at zero. Mount a 0-to-1 in. dial indicator on the intake pushrod of the number one cylinder as shown **(see illustration)**. The dial indicator shaft must be in exact alignment with the centerline of the pushrod. Preload the indicator stem about 0.100 in. and then reset the dial to zero. **Note:** *If the cylinder heads are not installed, mount the indicator so it presses against the lifter.*

Rotate the crankshaft clockwise until 0.050 in. movement occurs at the dial indicator. Note the position of the pointer in relation to the degree wheel. Record your reading and continue rotating the crankshaft until the lifter again returns to 0.050 in. lift. Record this reading.

Repeat the above procedure on the adjacent exhaust valve lifter and record the results. Compare the readings to the camshaft timing specification tag. If you want to be really thorough, check every cylinder - camshaft machining can be imperfect in some cases.

The readings you have taken may be two to four crankshaft degrees off of the timing tag figures. If this is the case, there are probably slight errors in the dowel pin hole location on the camshaft sprocket. These variations can be corrected by using an offset cam sprocket bushing. These are readily available in speed shops and from camshaft manufacturers. Follow the instructions provided with the kits. **Note:** *Some cam manufacturers build in one or two degrees of advance to compensate for timing chain wear.*

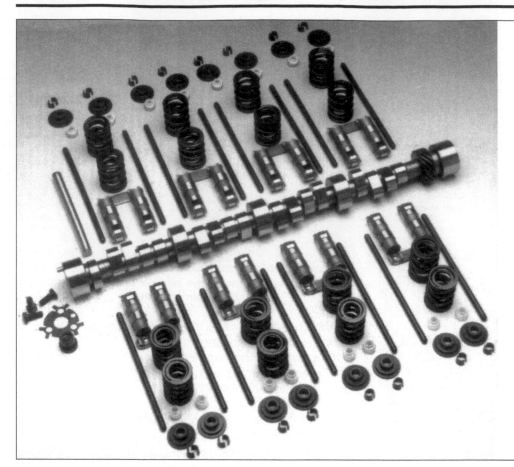

9.16 Buying a complete camshaft kit ensures compatibility of components

Matching valve train components

Most camshaft manufacturers sell camshaft kits **(see illustration)** that include the matching lifters, valve springs, pushrods, retainers and sometimes even the rocker arms. Purchasing all of the components from the same source, ensures compatibility of the parts. Also, if you have a problem, it's easier to deal with only one manufacturer.

Precautions

On hydraulic and mechanical lifter camshafts, don't install used lifters on a new camshaft. Used roller lifters may be installed on a new camshaft designed for roller lifters if the lifters are in good condition; however, it's always best to replace all components at the same time.

Apply a generous coating of assembly lube to all components prior to installation. Follow the manufacturer's instructions and don't take any short cuts. Let the cam and lifters break in, then change the oil frequently.

Rocker arms

Installing roller rocker arms **(see illustration)** is an easy way to gain a few horsepower and improve mileage by reducing friction. Roller rockers are available from many manufacturers and are compatible with all types of lifters. Buy a known quality brand and use quality hardware and pushrods to complete the installation.

Exhaust modifications

Stock manifolds

The original stock cast iron exhaust manifolds are quiet, sturdy, compact and you probably already have them. If you intend to modify your vehicle only slightly, they'll probably work fine.

9.17 Roller rocker arms reduce friction, add power and ensure more stable valve lash

Smog regulations in some areas prohibit the use of aftermarket tubular headers with catalytic converters. If these regulations apply to your vehicle, you can still pick up some additional power by adding dual exhaust while still retaining the original type manifolds.

Several good cast iron manifolds were used on high-performance and police models, especially in the 1960s. On small-block engines, the "ram's horn" design used in the 50's and 60's flows very well. Manifolds designed for Corvette are usually better than others. These may still be available from your dealer or a salvage yard. If you intend to use these on a different body style than they were originally designed for, be sure there is enough clearance and the outlet is in the right place.

Headers

Tubular steel exhaust headers generally provide increases in power without significant changes in fuel economy. They do this by allowing the exhaust gases to flow out of the engine more easily, thereby reducing pumping losses. Headers also "tune" or optimize the engine to run most efficiently in a certain rpm range. They do this by using the inertia of exhaust gases to produce a low-pressure area at the exhaust port just as the valve is opening. This phenomenon, known as scavenging, can be exploited by choosing headers of the correct type, length and diameter to match your engine type, driving style and vehicle.

Header manufacturers produce a large variety of headers designed for different purposes. Most of them provide detailed information, recommendations and specifications on their products to assist you in selecting the right headers for your application. Some companies also have technical hotlines to answer customer questions.

Types

Most headers come in a conventional four-into-one design (see illustration). This means that all four pipes on one side of the engine come together inside a large diameter collector pipe. These four-into-one designs tend to pro-

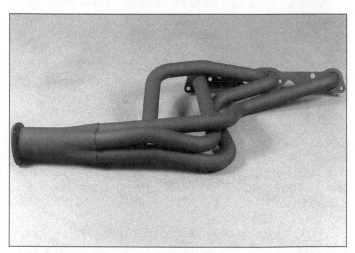

9.18 Four-into-one headers are the most common type

duce more peak power at high rpm at the expense of low and mid-range torque.

Another, less common approach to header design is commonly known as "Tri-Y" because of their appearance. The initial four pipes are paired into two pipes and then these are paired again into one pipe at the collector. This design provides more low and mid-range power, but sacrifices a slight amount of horsepower over about 6,000 rpm compared to the four-into-one design.

One of the more recent developments in exhaust system technology is the Anti-reversion (or AR) header. They have a small cone inside the pipe near the cylinder head that prevents reflected pressure waves from creating extra backpressure. The AR design also allows the use of larger diameter tubing without sacrificing a lot of low-end torque. Anti-reversion headers broaden and extend the torque curve and improve throttle response.

Length and diameter

Selection is not an exact science, but rather a series of compromises. Generally, long tubing and collector lengths favor low-end torque and shorter lengths produce their power in the higher rpm range. Larger diameter tubing should be used on large displacement engines and smaller cross-section tubing should be used on smaller displacement models.

Most street headers have primary tubes that are 30 to 40 inches long. Primary tube diameter should be about the same as exhaust port diameter. For small block engines this is about 1 5/8 in.. On big block engines, 1 7/8 in. is typical. The collectors are usually about 10 to 20 inches long and three to four inches in diameter.

Component matching

It's also important to match the characteristics of the intake and exhaust systems. Keep in mind that heavier vehicles with relatively small engines need more low-end torque and higher numerical gearing to launch them than lighter vehicles with larger engines. Follow the manufacturer's recommendations to ensure that you obtain the correct model for the application.

Open-plenum, single-plane intake manifolds with large, high-CFM carburetors should only be used with four-into-one headers designed for high rpm. This combination should also include a high-performance camshaft matched to the other components and a fairly high numerical gear ratio.

If you want a more tractable street engine, you may want to go for a 180-degree dual-plane manifold with a somewhat smaller carb and "Tri-Y" type headers or four-into-one headers with relatively long tubes. A mid-range camshaft would help complete the package.

Buying tips

Be sure to check the manufacturer's literature carefully to determine if the headers will fit your vehicle and allow you to retain accessories such as power steering, air conditioning, power brakes, smog pumps, etc. Note if there is sufficient clearance around the mounting bolts for a wrench. Also, some headers interfere with the clutch linkage on manual transmission models, so check for this, if applicable.

Some applications require the engine to be lifted slightly or even removed to allow header installation. Find out how the mufflers and tailpipes connect to the headers, also. Most installations require some cutting and modifications to the existing exhaust system, which may require a trip to the muffler shop. Check the instructions, ask questions and know what you are getting into before you purchase any components. Don't let poor planning get you stuck with a ticket for driving to the muffler shop with open headers!

Most headers don't come with heat shields like stock manifolds do and tend to melt original spark plug wires and boots. Plan on purchasing a set of heat-resistant silicone spark plug wires and boots, along with the necessary wire holders or looms to keep them away from the headers.

Also, look for header kits that have adapters for heat riser valves and automatic choke heat tubes. Some models with air pumps need threaded fittings in the headers to mount the air injection rails. If your vehicle has an exhaust gas oxygen sensor, be sure the replacement headers have a provision for mounting the sensor.

After the headers are installed, the fuel mixture may be too lean. Carbureted models usually require rejetting, and some fuel-injected engines may need fuel system adjustments. Test the vehicle on an engine analyzer after installation and tune the engine as necessary; failure to do so may result in driveability problems and/or burned valves.

Disadvantages

Headers have several disadvantages which should be considered before you run out and buy a set. Because of their thin tubing construction, headers emit more noise than cast-iron exhaust manifolds. Most headers produce a "tinny" sound in and around the engine compartment and some also produce a resonance inside the vehicle at certain engine speeds.

Tubular headers also have more surface area than cast-iron manifolds, so they give off more heat to the engine compartment. Engines with stock manifolds transmit this heat into the exhaust system instead.

Another caveat is port flange warpage. Some lower-quality headers have thin cylinder head port flanges which are more prone to warpage and exhaust leaks. Compare the thickness of the flanges on various brands and also check the gaskets and hardware supplied with the kits. The gaskets should be fairly thick and well made and the hardware should be of a high grade and fit your application. Look for the best design and quality control. Before you install headers, check the flanges for warpage with a straightedge, and if warpage is excessive, return them before they are used.

If you live in the "Rust Belt," be prepared for corrosion. Regular paint will quickly burn off and rusting will begin.

Before installation, thoroughly coat the headers with heat-resistant paint, porcelainizing or aluminizing spray.

Aftermarket tubular exhaust headers are not approved for use on catalytic converter-equipped vehicles in California and some other jurisdictions (engines equipped with tubular headers from the factory are approved). Additionally, vehicles originally equipped with air injection tubes in the manifolds must retain this system. Check the regulations in your area before you remove the original manifolds.

Exhaust systems

Substantial gains in both performance and economy throughout the rpm range can be obtained by properly modifying the stock exhaust system. The less restriction, the more efficient the engine will be. By increasing the flow capacity, you can unleash the potential horsepower in your engine.

The original mufflers, pipes and catalytic converters on most vehicles have small passages and tight bends. Perhaps the most difficult task when choosing exhaust components for a performance street vehicle is to make the system reasonably quiet while still having low restriction. Also, vehicles originally equipped with catalytic converters won't pass emissions inspections if the catalysts are removed.

Before you purchase a new exhaust system or individual components, you need to know what is available and the pros and cons of various items. Determine what you want your vehicle to be like, what your budget is, then design a balanced system that meets your requirements.

Many special parts and even complete high-performance systems are available for some popular models. If you have a less-common vehicle, you may have to devise a system composed of "universal" parts or use items originally designed for other vehicles.

If you intend to do the modifications on your vehicle yourself, find a level concrete area to work. Raise the vehicle with a floor jack and support it securely with sturdy jackstands. **Warning:** *Never work under a vehicle supported solely by a jack!*

Most exhaust system work requires only a small number of common hand tools. Sometimes, special equipment such as an acetylene torch, a pipe expander or pipe bender will be necessary. Muffler shops usually have this equipment and can modify, fabricate and install custom exhaust systems. But remember, the more work you do yourself, the more money you can save.

One of the easiest ways to reduce backpressure and increase flow capability is by adding dual exhaust. This effectively doubles the capacity of the exhaust without making the vehicle appreciably louder. Sometimes you can reuse most of the original parts on one side and install new parts on the other. If your vehicle is required to have a catalytic converter, you can install one on each side of the system to stay within the law.

As a general rule, mufflers and pipes with large inlet and outlet openings can handle more flow volume than smaller ones, all other factors being equal. Also, shorter

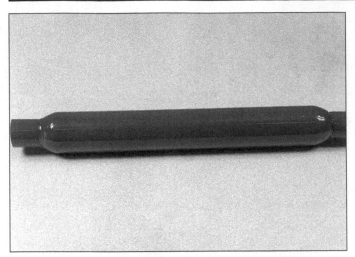

9.19 Glass pack mufflers use a "straight through" design - the larger diameter portion is packed with fiberglass or a similar material to help deaden noise

9.20 Turbo mufflers have large passages and high flow capacity

mufflers are usually louder and have less restriction than longer ones. So keep this in mind when choosing mufflers.

Conventional or original equipment mufflers

This is the type your vehicle came with from the factory. They are usually of reverse flow construction, and are quite heavy and restrictive. The only time you should consider using these is if you are adding dual exhaust and want to match the original muffler to save money.

Glasspack mufflers

Glasspack mufflers **(see illustration)** are usually less expensive than conventional or "turbo" mufflers, and have less restriction than original equipment types. Unfortunately, they're also too loud to meet most noise regulations and they vary greatly in quality and performance.

Turbo mufflers

Originally, factory engineers designed a large, high flow muffler for use on the turbocharged Chevrolet Corvair. Hot rodders found out and started using them for all sorts of vehicles. The name "turbo muffler" was first applied to these Corvair parts and gradually became generic, covering all high-flow oval shaped mufflers.

Turbo mufflers **(see illustration)** offer the best features of stock and glasspack mufflers. They are simply high-capacity mufflers that have low restriction without making much more noise than the original units. Today, virtually every muffler manufacturer sells a "turbo" muffler. They have expanded coverage to include many models; there should be one to fit your application.

Exhaust crossover pipes

Every dual exhaust system should have an exhaust crossover pipe located upstream of the mufflers. This tube balances the pulses between the two sides of the engine and allows excess pressure from one side to bleed off into the opposite muffler. This reduces noise and increases capacity. Be sure that any system you purchase incorporates this design, which improves power throughout the rpm range.

Stainless steel

Stainless steel mufflers and pipes are preferable to regular steel ones if you plan to keep your vehicle for a long time and/or do a lot of short-trip driving. Several specialty manufacturers produce ready-made and custom-built exhaust components from stainless steel. These parts are extremely resistant to corrosion and most are warranteed for the life of the vehicle. They usually cost several times more than standard components, so consider that when you make a purchase decision.

Pipe diameter

Exhaust pipe diameter is measured inside the pipe. Small increases in diameter result in large increases in flow capacity. As a general rule, small-block engines displacing 5.0 to 5.8 liters run well with 1 7/8 to 2 inch diameter exhaust pipes, with the more powerful ones using 2 1/4 inch. Engines over 6.5 liters almost always use 2 1/4 to 2 1/2 inch pipes.

Ignition systems

Pre-1975 model engines were equipped with point-type distributors. Standard models came with single-point distributors, while high-performance models often have dual-point versions. If your engine has a single-point distributor, it will probably benefit from an aftermarket performance-oriented replacement unit.

Breakerless electronic ignitions available in the after-

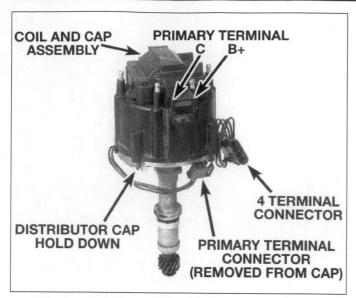

9.21 A typical Chevrolet electronic ignition distributor

9.22 A typical points-type ignition distributor

market and those installed by the factory since 1975 **(see illustration)** make conventional point-type distributors outmoded **(see illustration)**. However, all non-computer controlled ignitions can benefit from being checked and the advance curve prepped by an ignition shop on a distributor tester. If your engine has a high-performance dual-point distributor, you may want to retain it to keep the engine original. Stock distributors use an advance curve that is a compromise between economy, low-octane fuel and emissions. For most applications, total spark advance should not exceed 38-degrees. So if the centrifugal advance in the distributor produces 28-degrees advance at the crankshaft, initial timing should be about 10-degrees BTDC.

Be very careful when you vary significantly from original specifications. Too little advance results in lost power, overheating and reduced fuel economy. Too much advance can cause internal damage to the engine and high exhaust emissions. Be sure the timing is not advanced to the point where the engine pings continuously under load. A slight rattle when you accelerate is normal, but continued pinging will damage pistons, rods and bearings.

If you have a late-model vehicle with computer-controlled electronic ignition, about the only thing you can do is install an aftermarket control module. Some of these will cause a vehicle to fail an emissions test; check with the manufacturer of the product before you buy.

Some of the more recent model vehicles have knock sensors mounted in the block that detect any pinging and signal the ignition control module to retard the spark timing until the noise goes away. This feature also allows the ignition timing to advance for higher octane fuel and to retard if pinging is detected due to lower octane. Don't tamper with this system; it works well as it is.

Every ignition upgrade should include a new, high-quality distributor cap, rotor, spark plugs and ignition wires (and new points and condenser, if equipped). For street

machines, use radio interference suppression wires (TVRS); the solid non-suppression wires will wipe out radio reception (for you and cars and houses near you) and won't provide a measurable gain in performance. Use looms to keep the wires apart. Long parallel runs can cause induced crossfire.

A quality high-performance ignition coil is a nice extra. Be sure to get one that is compatible with the rest of the ignition system.

Intake manifolds

General information

A new intake manifold can unlock a considerable amount of power and improve economy at the same time if it is selected properly. Another benefit of using a high-performance manifold is the low weight of aluminum compared to the original cast-iron units.

Although exotic multi-carburetor high-rise and cross-ram manifolds look great, for street use, simple is better. Almost all current high-performance street manifolds are topped off with a single four-barrel carburetor. These provide power with economy and reliability at relatively low cost. Multiple carburetor setups are expensive to buy and difficult to tune and maintain. A single, properly selected carburetor will provide all the flow necessary for a street engine.

Intake manifolds, like many other parts of the engine, are "tuned" to perform best in a certain rpm range. Generally, manifolds with long runners produce more low rpm torque and manifolds with relatively short runners increase high rpm horsepower.

On carbureted and throttle body fuel-injected models, the intake manifold distributes the air/fuel mixture to the intake ports of the cylinder heads. Intake manifolds on port fuel-injected models carry only air; the fuel is injected into the ports.

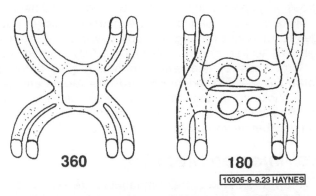

9.23 On 360-degree designs, all of the runners feed from a single chamber (or plenum) - on 180-degree designs, half the runners connect to one plenum, and half to the other

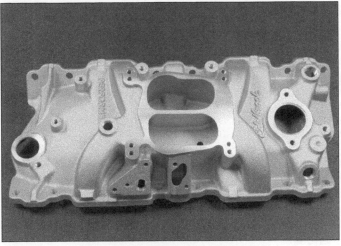

9.24 A typical dual-plane manifold

There are two basic intake manifold designs available for street machines, single plane and dual plane (also known as 360-degree and 180-degree, respectively).

Virtually all stock V8 intake manifolds use the dual-plane design because they enhance low and mid-range power, economy, driveability and produce low emissions. Dual plane-manifolds **(see illustrations)** are divided so every other cylinder in the firing order is fed from one side of the carburetor and the remaining cylinders are fed from the other side. This effectively improves intake velocity and throttle response at low and mid-range rpm at the expense of some top-end power.

On single-plane (360-degree) intake manifolds **(see illustration)**, all of the intake runners are on the same level and approximately the same length. This helps improve mixture distribution, which is a problem with some dual-plane manifolds. Recently, single-plane manifolds have been designed with better low and mid-range performance. If you want increased mid and top-end power instead of low and mid-range torque, a single-plane manifold may be the way to go. Keep in mind, though, that the most useable power

produced by an engine is in the low and mid rpm range.

Intake manifolds are available in low, mid and high-rise versions. The low-rise type is designed to fit under low hoodlines and generally sacrifices some power relative to the higher rise type. For most applications, if there is enough hood clearance, go with the high-rise type.

There are two other common designs that you should know about, which are used primarily for racing. Unfortunately, these manifolds decrease low-end performance, driveability and fuel economy and increase exhaust emissions.

The cross-ram manifold mounts two four-barrel carburetors side-by-side, instead of one in front of the other. Each carburetor feeds the cylinders on the opposite bank and provides a "ram-tuning" effect.

Tunnel-ram manifolds **(see illustration)** also mount two four-barrel carburetors, but they are installed inline instead of side-by-side. They look wild on "Pro-street" cars with the carburetors sticking through the hood, but don't work well on street-driven vehicles.

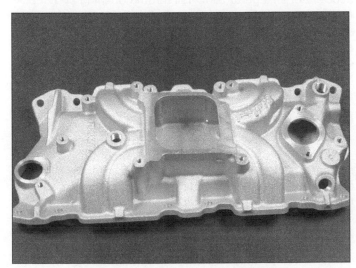

9.25 A typical single-plane manifold

9.26 A typical tunnel-ram manifold

Buying tips

There are a bewildering number of high-performance factory and aftermarket intake manifolds available for carbureted engines. Manufacturers provide a wealth of information in their catalogs to assist the buyer in making an informed purchase. However, the engines must be identical to the engines listed in the tests to produce the same amount of power. All of the components must work together for best results.

Once you decide on the basic type (single-plane or dual-plane) and rpm range, you must get the right manifold to fit your engine and carburetor. If the carburetor currently on the engine is in good condition and meets your needs, you can retain it and save money. If the carburetor is defective or doesn't suit the application, now is the time to change it.

The intake manifold mounting flanges for various carburetors are different, so you must decide what carb to use before you buy a manifold. For example, spread bore and regular bore carburetors have different bore sizes and bolt pattern spacing.

Many performance enthusiasts overcarburete their engines, which hurts throttle response, economy and emissions. The carburetor must be matched to the manifold and the engine for best results. See the Carburetor Section for further information.

Be sure to match the manifold characteristics with the overall theme and purpose of your vehicle. Manifold manufacturers spend countless hours testing their products on dynomometers. Some companies sell matched camshaft and manifold kits **(see illustration)**. Before you plunk down your hard-earned cash, find out from the manufacturer what the manifold is designed to do and what it will be like on your engine.

Emission-controlled models may be required to have a functional Exhaust Gas Recirculation (EGR) valve. Some aftermarket manifolds don't have a provision for EGR; this could cause you to fail an emissions test. Also, many

engines ping under load when the EGR is disconnected.

Make sure the manifold has exhaust crossover passages to improve warm-up driveability. Without these, the engine will hesitate and stumble until it is fully warmed up.

Sometimes replacement manifolds require different throttle linkage and mounting hardware. Some engines have a choke stove and many have numerous vacuum fittings. Be sure the parts are readily available in kit form or separately before you purchase a manifold.

Fuel-injected models

A few aftermarket intake manifolds are available for throttle body and port injected engines. Most of these are designed to improve power and economy while still remaining "smog" legal. Check local regulations before modifying any emission-controlled vehicle.

Installation tips

Always use new gaskets and seals during installation. Follow the instructions provided with the gasket set and the intake manifold and use quality, name-brand products.

If the cylinder heads have been resurfaced, be sure to have the machine shop mill the manifold surfaces to prevent leaks.

Aluminum manifolds are slightly more prone to warpage than their cast-iron counterparts. Follow the recommended tightening procedure, which is usually an alternating sequence from the center toward the ends, working from side-to-side. Use a torque wrench to tighten the fasteners to the torque specified by the manifold manufacturer.

Install a new thermostat and gasket whenever you replace the intake manifold. Use the correct heat range thermostat for the vehicle. Most emission-controlled models use a 195-degree F. thermostat. Earlier models generally use a 180-degree unit.

After the installation is complete, carefully adjust the accelerator linkage to allow full throttle opening and check for binding and sticking before starting the engine.

Be sure there is sufficient fresh oil and coolant in the engine. Run the engine, set the ignition timing, adjust the carburetor (if equipped) and check for oil, fuel and coolant leaks.

Carburetors

There is probably more conflicting information and "old mechanic's tales" about carburetors than any other part of a vehicle. Carburetors simply mix fuel and air and control the amount of this fuel/air mixture going into the engine at any given time. However, the way they perform these simple-sounding functions can get quite complicated, especially on emission-controlled vehicles.

It helps to know a little about carburetor basics. Despite common belief, engines don't really suck fuel out of

9.27 Some companies make kits with compatible camshafts and intake manifolds

the carburetor. All carburetors have a venturi, which is a narrowing of the throat. As air rushes through this restriction, a pressure drop occurs. A small orifice is installed at this point for fuel delivery. Atmospheric pressure acting on top of the fuel forces the fuel from the carburetor bowl, through the orifice and into the carburetor throat, where it enters the intake manifold and then the cylinder.

Engines require different air/fuel ratios (or mixtures) when the engine is cold, warmed up, at idle, at cruise and under heavy load. Carburetors have a number of systems to help it function under various operating conditions. In addition to the systems discussed below, there are many add-on components such as anti-dieseling solenoids and dashpots that are used on specific applications. These items were put there for a reason, and removing them can have a detrimental effect on driveability.

Float system

The float system maintains the level of fuel in the carburetor bowl. It works like the float in a toilet tank. As the level goes down, the float drops, which opens the needle valve and allows more fuel into the carb. By keeping the fuel level within a narrow range, the fuel/air ratio can be controlled more accurately. For best performance, the float level must be adjusted precisely to factory specifications.

Choke

The choke system enables the engine to start when it's cold by enriching the fuel mixture. What it does is "choke" off the supply of air so the engine gets proportionately more fuel. When this happens, the idle speed is reduced, so a fast-idle system is added to the throttle linkage to boost the idle speed during warm-up. No modifications of this system are necessary for a street machine.

Idle system

The idle system provides the fuel necessary to keep the engine running at low speeds, when the main metering system is ineffective. Adjustment screws allow changes in the air/fuel ratio at idle (on many emission-controlled models, the screws are covered by tamper-resistant plugs). Many home mechanics are under the impression this adjustment changes the fuel mixture throughout the rpm range; it does not.

Accelerator pump

The accelerator pump provides an extra squirt of fuel when the throttle is opened quickly to prevent the engine from hesitating or backfiring through the intake during acceleration. If you look down the throat of the carburetor (with the engine off) and move the throttle linkage quickly, fuel should squirt from the pump discharge orifices.

A mechanical linkage connects the accelerator pump to the throttle linkage on the carburetor. This and the pump stroke are adjustable on most carburetors. If the engine seems to bog during initial acceleration, check the operation of the accelerator pump.

Off-idle system

The off-idle system provides a transition between the idle and main metering systems. Many carburetors have transfer slots or ports near the throttle valve plates that feed fuel as they are uncovered during throttle opening.

Main metering system

The main metering system meters fuel to the engine at cruise speeds. It's made up of a main jet, main nozzle and venturi. The main jet is located in the passage between the carburetor bowl and the main nozzle. The main nozzle usually consists of a tube with small air bleed holes. The air and fuel mix here to form a mist of atomized fuel.

The main jet determines how much fuel will be mixed with a given quantity of air. Tuners use different size main jets to tailor a carburetor to an engine and its operating environment. By using a larger jet, the fuel mixture is enriched. Conversely, a smaller jet leans out the mixture. An engine run at high altitude needs smaller jets than the same engine would at sea level.

Full-load enrichment

Engines need a richer fuel mixture when they are under a heavy load than they need at cruise. The full-load enrichment system provides additional fuel when the engine is under heavy loads and full throttle.

Several types of full-load enrichment systems are used in different brands of carburetors. The most common are the diaphragm-operated power valve, metering rods and power bypass jet or enrichment valve.

Diaphragm-operated power valves are the type found on Holley and some other carburetors. When intake manifold vacuum reaches a certain preset point, the valve opens, letting additional fuel into the engine. Some models have two-stage valves to provide more precise metering. The power valves are rated according to their opening point, measured in inches of mercury manifold vacuum. Tuners can match the power valve to the application. Engines that normally develop low vacuum (ones with hot cams, etc.) should be equipped with power valves that open at lower vacuum readings.

Metering rods move in and out of metering orifices (usually the main jets) in reaction to intake manifold vacuum. When the engine is under heavy load and vacuum drops, the rods slide out of the main jets to increase fuel delivery.

Power bypass jets perform the same function as a metering rod, except they have their own jet or enrichment valve (sometimes called an economizer valve).

General information

If your vehicle originally came with one or more carburetors, you need to consider all the options before you throw the old fuel mixer(s) in the trash. If you're doing a restoration, the original carburetor(s) must stay.

Factory triple two-barrel and dual-quad setups increase the value of collector vehicles; never discard such

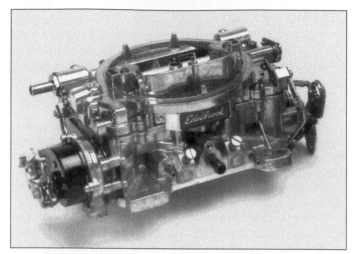

9.28 Conventional four-barrel carburetors, such as the Edelbrock/Carter AFB model, work well in most street applications

9.29 On spread-bore carburetors, the primary barrels are much smaller than the secondaries

9.30 Dual-feed carburetors are identified by the two fuel inlets (arrows)

systems. A careful rebuild and rejetting is about all you can do. If you do decide to install an aftermarket carb, save the old setup for when you sell the vehicle.

Emission controlled vehicles present a special problem. If your vehicle is subject to smog inspection, you have to use the original type or an approved replacement. Later models equipped with exhaust gas oxygen sensors are especially difficult to replace. Check your state's requirements before changing any emission-related components.

There are so many choices available in aftermarket carburetors that it boggles the mind. To make matters more complicated, retrofit fuel injection systems are becoming increasingly popular. However, even the least expensive fuel-injection conversion costs several times as much as a good carburetor. If you can afford fuel injection, it's the way to go.

Most high-performance V8-powered street vehicles use four-barrel carburetors, and we will limit our discussion to them. They provide the most "bang for the buck" of any system available.

Four-barrel carbs provide the best of both worlds. During light throttle and cruising conditions, the engine runs on the front two barrels. This keeps the air flow velocity relatively high through the carburetor for optimum fuel mixing. As the accelerator is mashed to the floor, the rear two barrels are opened, which transforms the carburetor into a high-capacity unit. There are several basic types of popular four-barrel carbs:

Conventional four-barrel carburetors (see illustration) have the primary and secondary throttle bore sizes approximately equal. These carbs are very common and work well in most applications.

Spread bore carburetors (see illustration) are designed to exploit the progressive characteristic of four-barrel carburetors. The front (primary) barrels are substantially smaller than the rear (secondary) ones, so off-idle and low-speed economy are improved. When the rear barrels

open, gobs of additional power are produced (and gobs of fuel are used).

"Double-pumper" carburetors have a separate accelerator pump on the primary and secondary barrels. This eliminates any "flat spots" on acceleration, but increases fuel consumption and emissions.

Dual-feed carburetors (see illustration) by Holley have two fuel inlet fittings, one on each float bowl. This design is a carry-over from racing, where large quantities of fuel must be delivered to the engine. It works well on large-displacement and high-horsepower street engines, but is overkill on most small-block engines.

Carburetor sizing

Matching the carburetor to the engine is critical for performance and economy. Many hot rodders like the look of a huge carburetor on their engine and succumb to the temptation of "more is better".

If an engine is overcarbureted, it will bog and hesitate

at low speeds and won't start to run well until very high rpm. Fuel economy and emissions will be poor.

Larger displacement engines and engines that run at high rpm need larger capacity carburetors than smaller engines running at lower speeds.

Many carburetors are rated according to their potential air-flow capacity in Cubic Feet per Minute (CFM). Most, but not all manufacturers test their carbs at 1.5 inches of mercury (in-Hg); when you compare different models, be sure both are measured the same way.

The most important factors in carburetor size selection are engine displacement, maximum rpm and volumetric efficiency.

Volumetric efficiency is a measure of the engines ability to fill the cylinder completely, and is given as a percentage. For example, a 100 cubic inch engine that gets 80 cubic inches of air/fuel mix into the combustion chamber on each intake stroke has a volumetric efficiency of 80%.

For the sake of simplicity, we have assumed a volumetric efficiency of about 80 percent, which is average for a four-barrel equipped high-performance street engine. You must decide what rpm range you want your engine to run best at. Be realistic - you're only hurting yourself by overestimating. Round off all results to the nearest carburetor size. Refer to the accompanying chart for guidelines on carburetor capacity selection.

When selecting a carburetor, use this chart to estimate the required CFM rating.

Engine RPM

Engine displacement (cu. in.)	4000	4500	5000	5500	6000	6500
250	245	260	290	320	350	380
275	255	290	320	350	380	420
300	280	315	350	380	420	450
325	300	340	380	415	450	490
350	325	365	405	445	490	525
375	350	390	435	480	520	565
400	370	420	465	510	555	600
425	400	450	500	550	600	650
450	420	470	520	580	625	700

Note: *This chart is for stock street engines. For high-performance models, increase the CFM rating about ten percent.*

As a general rule, high-performance small-block engines will need 500 to 600 CFM carbs, depending on actual displacement and level of modifications. Performance big-block engines work well with 650 to 800 CFM carbs, again depending on displacement and level of modifications.

Be sure to get a carburetor that will fit on your manifold and can be adapted to the factory throttle linkage. If the carb has vacuum secondaries, leave them intact - don't try to convert them to mechanical operation. Also, the original air cleaner may not fit on a different carb. On pre-smog models, get a high-flow filter to go with the new carburetor **(see illustration)**. If your vehicle is emission controlled, get an approved replacement carburetor and air filter assembly and reconnect all smog controls. After a new carburetor is installed, have it adjusted on an exhaust gas analyzer for optimum efficiency.

9.31 Aftermarket high-flow air cleaners are available to fit most four-barrel carburetors - some are available in low-profile designs to reduce the possibility of hood interference

Most carbureted engines have mechanical fuel pumps. If you modify the engine substantially, install a high-volume mechanical pump or mount an electric fuel pump near the gas tank. Follow the instructions provided with the pump to assure a safe installation.

Making it all work together

After you have performed all of the modifications and installed all of the right parts, the moment of truth arrives. Does the vehicle run any better than it did before?

If you've just overhauled the engine, break it in carefully (see Chapter 7) before attempting any performance tests. Then carefully tune the engine to extract its full potential. The best way to do this is to run it on a chassis dynamometer while the engine is connected to a diagnostic scope and emissions analyzer. This insures that everything is up to snuff and allows you to make accurate jetting and timing changes.

Dyno tuning is one of the best investments you can make in your vehicle. Frequently, up to 50 horsepower are found simply by tuning and making simple adjustments.

There are also some changes you can make to reduce friction losses. Try using special low-friction and synthetic lubricants. Clutch fans use less power than solid hub fans. Bias-ply tires have more rolling resistance than radials. when the old tires wear out, install a set of performance radials (don't mix bias ply and radial tires on the same vehicle).

Engine swaps

Sometimes you just can't make the original engine in your vehicle perform the way you want, no matter how it's modified. You like the vehicle otherwise, but the engine just doesn't hack it. Perhaps it's time for an engine swap.

Engine swapping opens the door to an almost unlimited variety of engines. However, there are practical considerations that limit your choices. First, the engine must be physically small enough to fit in the engine compartment. This isn't a big problem on full-size cars and trucks, but

prevents you from putting big-block engines into sub-compacts, etc.

The easiest swaps are between engines from the same family (big block or small block). For example, if you decide to swap your tired low-compression two-barrel carbureted 327 for a freshly rebuilt 327 high-performance version, it's basically a bolt-in job. Similarly, replacing a worn-out 396 with a 454 consists mostly of removing the old engine and installing the new one.

Swapping engines from different families within the same brand is the next easiest way to go. Most engine swappers take out a small engine and put in a larger and/or more powerful one that was optional in that model. For instance, they'll take out the original ailing six-cylinder engine and install an optional V8. The best way to do this economically is to purchase a wrecked vehicle similar to your own, with a complete intact drivetrain. Then swap all the parts that are different and scrap what's left over. That way, you have all the parts and wiring you need and can see exactly how everything is installed.

The most difficult type of swap is to install an engine from another brand of vehicle into a chassis. If it's a popular swap, there may be aftermarket conversion kits that help considerably. Check very carefully before you attempt such a conversion; countless thousands of such half-finished projects are languishing in garages and yards around the country.

Before you begin a project, there are several items to consider. If the new engine is considerably heavier or lighter than the old unit, ride and handling will be affected. Sometimes, changing the springs will solve the problem, but if there is too much weight in the front, the vehicle will not handle safely.

Another area of concern is emission regulations. If your vehicle is subject to smog inspections, certain changes are prohibited. For example, in California, you can put a newer engine in an old chassis, but not the opposite. Also, the emissions systems for the newer engine must be intact and operative.

Engine swaps can be very beneficial if they are planned and executed properly, but be sure to check the latest regulations in your area before you start.

Glossary

A

ABDC - After Bottom Dead Center.

Adjustable rocker arm - A type of rocker arm with an adjusting nut that can be tightened or loosened to adjust valve lash.

ATDC - After Top Dead Center.

B

Babbit bearing - A plain bearing made of babbit. Babbit is the trademark for a soft, silvery antifriction alloy composed of tin with small amounts of copper and antimony.

Backlash - The amount of "play" between two parts. Usually refers to how much one gear can be moved back and forth without moving gear with which it's meshed.

Bank - The portion of a V8 block containing four in-line cylinders.

Base circle - The part of the cam with the smallest diameter from the camshaft center. The area of the cam directly opposite the nose (lobe). No lift is produced by the base circle.

BBDC - Before Bottom Dead Center.

Bearing caps - The caps held in place by nuts or bolts which, in turn, hold the bearing halves in place.

Bearing clearance - The amount of space left between shaft and bearing surface. This space is for lubricating oil to enter.

Bearing crush - The additional height which is purposely manufactured into each bearing half to ensure complete contact of the bearing back with the housing bore when the engine is assembled.

Bearing knock - The noise created by movement of a part in a loose or worn bearing.

Bearing scraper - A small, triangular tool that looks like a file without teeth. Used for deburring and chamfering the edges of camshaft bearings.

Bearing spin - A type of bearing failure in which a lack of lubrication overheats the bearing until it seizes on the shaft, shears its locking lip and rotates in the housing or block.

Bearing spread - A purposely manufactured small extra distance across the parting faces of the bearing half, in excess of the actual diameter of the housing bore.

Bearing tang - A notch or lip on a bearing shell used to correctly locate the bearing during assembly.

Bellhousing (clutch housing) - The metal shroud which covers the flywheel and clutch, or the torque converter assembly.

Big-end bearing - The bearing in the end of the connecting rod that attaches to the crankshaft.

Bleed down - The collapse of a hydraulic lifter as oil drains out.

Block - The lower part of the engine containing the cylinders; the block is the basic framework of the engine.

Block deck - The cylinder head gasket surface. Blow-by - Leakage of compressed air-fuel mixture and burned gases from the combustion chamber past the piston rings into the crankcase. This leakage results in power loss and oil contamination.

Blueprinting - Dismantling an engine and reassembling it to EXACT specifications.

Bore - An engine cylinder, or any cylindrical hole; also used to describe the process of enlarging or accurately refinishing a hole with a cutting tool, as "to bore an engine cylinder." The bore size is the diameter of the hole.

Bore diameter - Diameter of the cylinders.

Boring - Renewing the cylinders by cutting them out to a specified size. A boring bar is used to make the cut.

Boring bar - An electric-motor powered cutting tool used to machine, or bore, an engine cylinder, thereby removing metal and enlarging the cylinder diameter.

Boss - An extension or strengthened section that holds the end of a pin or a shaft; for example, the projections within a piston for supporting the piston pin or piston-pin bushings.

Bottom end - A term which refers collectively to the engine block, crankshaft, main bearings and the big ends of the connecting rods.

Bounce - A condition in which the valve isn't held tightly closed against its seat, even though the camshaft isn't opening it.

Break-in - The period of operation between installation of new or rebuilt parts and time in which parts are worn to the correct fit. Driving at reduced and varying speed for a specified mileage to permit parts to wear to the correct fit.

Breathing - Air flow into an engine during operation.

BTDC - Before Top Dead Center; any position of the piston between bottom dead center and top dead center, on the upward stroke.

Build date code - A code which tells you what day, month and year the engine was made. Expressed alphanumerically and stamped somewhere on the block.

Burnishing - A sizing process that pushes metal to size by pressure.

Burr - A rough edge or area remaining on metal after it has been cast, cut or drilled.

Bushing - A one-piece sleeve placed in a bore to serve as a bearing surface for shaft, piston pin, etc. Usually replaceable.

Butt - The square ends of a piston ring.

C

Calibrate - To check or correct the initial setting of a test instrument; as in adjusting the needle of a dial gauge to correct zero or load setting.

Caliper - A measuring tool that can be set to measure inside or outside dimensions of an object; used for measuring things like the thickness of a block, the diameter of a shaft or the bore of a hole (inside caliper).

Cam - A rotating lobe or eccentric which, when used with a cam follower, can change rotary motion to reciprocating motion.

Cam follower - A device that follows the cam contour as it rotates. Also called a lifter, valve lifter or tappet.

Cam profile - The shape of the cam lobe.

Camshaft - The shaft in the engine, on which a series of lobes are located for operating the valve mechanisms. The camshaft is driven by gears or sprockets and a timing chain. Usually referred to simply as the cam.

Camshaft gear - The gear used to drive the camshaft.

Carbon - Hard, or soft, black deposits found in combustion chamber, on plugs, under rings, on and under valve heads.

Case-hardened - A piece of steel that has had its outer surface hardened while the inner portion remains relatively soft. Camshaft lobes are often case hardened.

Casting number - The number cast into a block, head or other component when the part is cast. Casting numbers can be helpful when identifying an engine or its parts, but they're not completely accurate, because casting are sometimes machined differently.

Cast iron - An alloy of iron and more than two percent carbon, used for engine blocks and heads because it's relatively inexpensive and easy to mold into complex shapes.

Center line - An imaginary line drawn lengthwise through center of an object. Chamfer - To bevel across (or a bevel on) the sharp edge of an object.

Chase - To repair damaged threads with a tap or die.

CID - Cubic Inch Displacement.

Clearance - The amount of space between two parts. For example, between piston and cylinder, bearing and journal, etc.

Coil binding - Compressing a valve spring to the point at which each coil touches the adjacent coil. See "Solid height."

Cold lash - The valve lash clearance, measured between the rocker arm and valve tip, when the engine is cold.

Collapsed piston - A piston whose skirt diameter has been reduced by heat and the forces imposed upon it during service in engine.

Combustion chamber - The space between the piston and the cylinder head, with the piston at top dead center, in which air-fuel mixture is burned.

Combustion chamber volume - The volume of the combustion chamber (space above piston with piston on TDC) measured in cc (cubic centimeters).

Compression height - The distance from the wrist-pin-bore center to the top of the piston.

Compression ratio - The relationship between cylinder volume (clearance volume) when the piston is at top dead center and cylinder volume when the piston is at bottom dead center.

Compression ring - The upper ring, or rings, on a piston, designed to hold the compression in the combustion chamber and prevent blow-by.

Compression stroke - The piston's movement from bottom dead center to top dead center immediately following the intake stroke, during which both the intake and exhaust valves are closed while the air-fuel mixture in the cylinder is compressed.

Connecting rod - The rod that connects the crank on the crankshaft with the piston. Sometimes called a con rod.

Connecting rod bearing - See "Rod bearing."

Connecting rod cap - The part of the connecting rod assembly that attaches the rod to the crankpin.

Core plug - Soft metal plug used to plug the casting holes for the coolant passages in the block.

Counterbalancing - Additional weight placed at the crankshaft vibration damper and/or flywheel to balance the crankshaft.

Counterbore - A concentric machined surface around a hole opening.

Crankcase - The lower part of the engine in which the crankshaft rotates; includes the lower section of the cylinder block and the oil pan.

Crank kit - A reground or reconditioned crankshaft and new main and connecting rod bearings.

Crankpin - The part of a crankshaft to which a connecting rod is attached.

Crankshaft - The main rotating member, or shaft, running the length of the crankcase, with offset "throws" to which the connecting rods are attached; changes the reciprocating motion of the pistons into rotating motion.

Crankshaft gear - The gear on the front of the crankshaft which drives the camshaft gear.

Crankshaft runout - A term used to describe how much a crankshaft is bent.

Crank throw - One crankpin with its two webs.

Creep - When a crankshaft has slightly excessive

runout (is slightly bent), it can sometimes be corrected by laying the crank in its saddles, installing the center main bearing cap (with its bearing insert) and leaving it for a day or two. Sometimes the crank will "creep," or bend, enough to put it within the specified runout range.

Cross-hatch - The pattern created on the cylinder wall by a hone.

Crush - A slight distortion of the bearing shell that holds it in place as the engine operates.

Cu. In. (C.I.) - Cubic inches.

Cubic inch displacement - The cylinder volume swept out (displaced) by all of the pistons in an engine as the crankshaft makes one complete revolution.

Cycle (four-stroke) - A repetitive sequence of events. In an engine, the term refers to the intake, compression, power and exhaust strokes that take place during each crankshaft revolution.

Cylinder sleeve - A replaceable sleeve, or liner, pressed into the cylinder block to form the cylinder bore.

Cylinder surfacing hone - Puts a cross-hatch pattern on the cylinder walls, after they've been bored, to help seat the new rings properly.

D

Damper - See Vibration damper.

Dead center - A term which refers to the maximum upper or lower piston position when all movement in one direction stops before it reverses direction.

Deburring - Removing the burrs (rough edges or areas) from a bearing.

Decarbonizing - The process of removing carbon from parts during overhaul.

Deck - The flat upper surface of the engine block where the head mounts.

Deck height - The center of the crankshaft main-bearing bores to the block deck surface.

Deglazer - A tool, rotated by an electric motor, used to remove glaze from cylinder walls so a new set of rings will seat.

Degreasing - The process of removing grease from parts.

Degree wheel - A disc divided into 360 equal parts that can be attached to a shaft to measure angle of rotation.

Detonation - In the combustion chamber, an uncontrolled second explosion (after the spark occurs at the spark plug) with spontaneous combustion of the remaining compressed air-fuel mixture, resulting in a pinging noise. Commonly referred to as spark knock or ping.

Dial indicator - A precision measuring instrument that indicates movement to a thousandth of an inch with a needle sweeping around dial face.

Die - see "Thread die."

Dieseling - The tendency of an engine to continue running after the ignition is turned off. Usually caused by buildup of carbon deposits in the combustion chamber.

Dish - A depression in the top of a piston.

Displacement - The total volume of air-fuel mixture an engine is theoretically capable of drawing into all cylinders during one operating cycle. Also refers to the volume swept out by the piston as it moves from bottom dead center to top dead center.

Dome - See "Pop-up."

Dowel pin - A steel pin pressed into shallow bores in two adjacent parts to provide proper alignment.

Draw file - Smoothing a surface with a file moved sidewise.

Dry manifold - An intake manifold with no integral coolant passages cast into it.

Dwell - The number of degrees that the breaker cam (on the distributor shaft) rotates from the time the breaker points close until they open again.

Dwell meter - A precision electrical instrument used to measure the cam angle, or dwell, or number of degrees the distributor points are closed while the engine is running.

Dykem-type metal bluing - A special dye used to check a valve job. When applied to the valve seat to show up as a dark ring contrasted against the brightly finished top and bottom cuts, making the seat easier to see and measure.

Dynamic balance - The balance of an object when it's in motion. When the center line of weight mass of a revolving object is in the same plane as the center line of the object, that object is in dynamic balance.

Dynamometer - A device for measuring the power output, or brake horsepower, of an engine. An engine dynamometer measures the power output at the flywheel.

E

Eccentric - A disk, or offset section (of a shaft, for example) used to convert rotary motion to reciprocating motion. Sometimes called a cam.

Edge-ride - The tendency of crankshaft main bearings to ride up the radius - rather than seat on the journal - when the radius is too large.

Endplay - The amount of lengthwise movement between two parts. As applied to a crankshaft, the distance that the crankshaft can move forward and back in the cylinder block.

Engine tune-up - A procedure for inspecting, testing and adjusting an engine, and replacing any worn parts, to restore the engine to its best performance.

Erode - Wear away by high velocity abrasive materials.

Exhaust manifold - A part with several passages through which exhaust gases leave the engine combustion chamber and enter the exhaust pipe.

Exhaust stroke - The portion of the piston's movement devoted to expelling burned gases from the cylinder. The exhaust stroke lasts from bottom dead center to top dead center, immediately following the power stroke, during which the exhaust valve opens so the exhaust gases can escape from the cylinder to the exhaust manifold.

Exhaust valve - The valve through which the burned air-fuel charge passes on its way from the cylinder to the exhaust manifold during the exhaust stroke.

Externally balanced crankshaft - A crankshaft that requires external balancing weight - usually on the vibration damper or the flywheel - for balance.

F

Face - A machinist's term that refers to removing

metal from the end of a shaft or the "face" of a larger part, such as a flywheel.

Fatigue - A breakdown of material through a large number of loading and unloading cycles. The first signs are cracks followed shortly by breaks.

Feeler gauge - A thin strip of hardened steel, ground to an exact thickness, used to check clearances between parts.

Finishing stone - see "Hone."

Firing order - The order in which the engine cylinders fire, or deliver their power strokes, beginning with the number one cylinder.

Float - Float occurs when the valve train loses contact with the cam lobe and the parts "float" on air until control is regained.

Flywheel - A heavy metal wheel that's attached to the crankshaft to smooth out firing impulses. It provides inertia to keep crankshaft turning smoothly during periods when no power is being applied. It also serves as part of the clutch and engine cranking systems.

Flywheel ring gear - A gear, fitted to the outer circumference of the flywheel, that's engaged by the teeth on the starter motor pinion (drive gear) to start the engine.

Foot-pound - A unit of measurement for work, equal to lifting one pound one foot.

Foot-pound (tightening) - A unit of measurement for torque, equal to one pound of pull one foot from the center of the object being tightened.

Four-stroke cycle - The four piston strokes - intake, compression, power and exhaust - that make up the complete cycle of events in the four-stroke cycle engine. Also referred to as four-cycle and four-stroke.

Free height - The unloaded length or height of a spring.

Freeplay - The looseness in a linkage, or an assembly of parts, between the initial application of force and actual movement. Usually perceived as "slop" or slight delay.

Freeze plug - See "Core plug."

Fulcrum - The hinge point of a lever. On a fulcrum-type rocker arm, the part on which the rocker arm pivots.

G

Gallery - A large passage in the block that forms a reservoir for engine oil pressure.

Glaze - The very smooth, glassy finish that develops on cylinder walls while an engine is in service.

Glaze breaker - See "Deglazer."

H

Half-round file - A special file that's flat on one side and convex on the other.

Hardened pushrods - Specially treated pushrods designed for use with pushrod-guided rocker arms.

Harmonic balancer - See "Vibration damper."

Head - See cylinder head.

Headers - High-performance exhaust manifolds that replace the stock manifold. Designed with smooth flowing lines to prevent back pressure caused by sharp bends, rough castings, etc.

Heat checks - Cracks in the clutch pressure plate.

Heli-Coil - A rethreading device used when threads are worn or damaged. The device is installed in a retapped hole to reduce the thread size to the original size.

Horsepower - A measure of mechanical power, or the rate at which work is done. One horsepower equals 33,000 ft-lbs of work per minute. It's the amount of power necessary to raise 33,000 pounds a distance of one foot in one minute.

Hot lash - The valve adjustment on an engine equipped with solid lifters.

Hydraulic valve lifter - Valve lifter that utilizes hydraulic pressure from engine's lubrication system to maintain zero clearance (keep it in constant contact with both camshaft and valve stem). Automatically adjusts to variation in valve stem length. Hydraulic lifters reduce valve noise.

I

Idle speed - The speed, or rpm, at which the engine runs freely with no power or load being transferred when the accelerator pedal is released.

Installed height - The spring's measured length or height, as installed on the cylinder head. Installed height is measured from the spring seat to the underside of the spring retainer.

Intake manifold - A part with several passages through which the air-fuel mixture flows from the carburetor to the port openings in the cylinder head.

Intake stroke - The portion of the piston's movement, between top dead center and bottom dead center, devoted to drawing fuel mixture into engine cylinder. The intake stroke is the stroke immediately following the exhaust stroke, during which the intake valve opens and the cylinder fills with air-fuel mixture from the intake manifold.

Intake valve - The valve through which the air-fuel mixture is admitted to the cylinder.

J

Jam nut - A nut used to lock an adjustment nut, or locknut, in place. For example, a jam nut is employed to keep the adjusting nut on the rocker arm in position.

Journal - The surface of a rotating shaft which turns in a bearing.

K

Keeper - The split lock that holds the valve spring retainer in position on the valve stem.

Key - A small piece of metal inserted into matching grooves machined into two parts fitted together - such as a gear pressed onto a shaft - which prevents slippage between the two parts.

Keyed - Prevented from rotating with a small metal device called a key.

Keyway - A slot cut in a shaft, pulley hub, etc. A square key is placed in the slot and engages a similar keyway in the mating piece.

Knock - The heavy metallic engine sound, produced in the combustion chamber as a result of abnormal combustion - usually detonation. Knock is usually caused by a loose or worn bearing. Also referred to as detonation, pinging and spark knock. Connecting rod or main bearing knocks are created by too much oil clearance or insufficient lubrication.

Knurl - A roughened surface caused by a sharp wheel that displaces metal outward as its sharp edges push into the metal surface.

L

Lands - The portions of metal between the piston ring grooves.

Lapping the valves - Grinding a valve face and its seat together with lapping compound.

Lash - The amount of free motion in a gear train, between gears, or in a mechanical assembly, that occurs before movement can begin. Usually refers to the lash in a valve train.

Lifter - The part that rides against the cam to transfer motion to the rest of the valve train.

Lifter foot - The part of the lifter that contacts the camshaft.

Load at installed height - The specified range of force required to compress a spring to its installed height, usually expressed in terms of so many pounds of force at so many inches.

Lug - To operate an engine with high loads at low speeds.

Lugs - The heavy fastening flanges on parts.

M

Machining - The process of using a machine to remove metal from a metal part.

Main bearings - The plain, or babbit, bearings that support the crankshaft.

Main bearing caps - The cast iron caps, bolted to the bottom of the block, that support the main bearings.

Manifold - A part with several inlet or outlet passages through which a gas or liquid is gathered or distributed. See "Exhaust manifold" and "Intake manifold."

Manifold runners - A single passage in a manifold from one cylinder to the major manifold opening.

Manifold vacuum - The vacuum in the intake manifold that develops as a result of the vacuum in the cylinders on their intake strokes.

Metering orifice - A small hole that restricts the flow of liquid - usually coolant or oil.

Micrometer - A precision measuring instrument which can measure the inside or outside diameter of a part to a ten-thousandth of an inch.

Mike - See "Micrometer."

Mill - To remove metal with a milling machine.

N

Notched rocker arm stud - A rocker arm stud with a notch worn in its side. A notched stud is more likely to break.

O

O.D. - Outside diameter.

OHV - Overhead Valve.

Oil gallery - A pipe or drilled passageway in the engine used to carry engine oil from one area to another.

Oil pan - The detachable lower part of the engine, made of stamped steel, which encloses the crankcase and acts as an oil reservoir.

Oil pumping - Leakage of oil past the piston rings and into the combustion chamber, usually as a result of defective rings or worn cylinder walls.

Oil ring - The lower ring, or rings, of a piston; designed to prevent excessive amounts of oil from working up the cylinder walls and into the combustion chamber. Also called an oil-control ring.

Oil seal - A seal which keeps oil from leaking out of a compartment. Usually refers to a dynamic seal around a rotating shaft or other moving part.

Oil slinger - A cone-shaped collar that hurls oil back to its source to prevent leakage along a shaft.

Orifice - See "Metering orifice."

O-ring - A type of sealing ring made of a special rubberlike material; in use, the O-ring is compressed into a groove to provide the sealing action.

Out-of-round journal - An oval or egg-shaped bearing journal.

Overhaul - To completely disassemble a unit, clean and inspect all parts, reassemble it with the original or new parts and make all adjustments necessary for proper operation.

P

Pedestal pivot - A semi-cylindrical (half-round) pivot used with pivot-guided rocker arms. A pedestal pivot restricts the rocker arm so it pivots around one axis or in a single plane - the plane of the valve stem and pushrod.

Penetrating oil - Special oil used to free rusted parts so they can be moved.

Phosphate coating - A special coating on camshafts which promotes oil retention.

Pilot - A term which refers to any device used to center a cutting tool or the concentric installation of one part onto another. For example, the valve guide is used as a pilot for centering the grinding stone during a valve job; the pilot bearing is used to center a clutch alignment tool when bolting the clutch disc and pressure plate to the flywheel.

Pilot bearing - A small bearing installed in the center of the flywheel (or the rear end of the crankshaft) to support the front end of the input shaft of the transmission.

Pinning - A procedure for repairing cracks in the combustion chamber using threaded pins.

Pip mark - A little dot or indentation which indicates the top side of a compression ring.

Piston - The cylindrical part, attached to the connecting rod, that moves up and down in the cylinder as the crankshaft rotates. When the fuel charge is fired, the piston transfers the force of the explosion to the connecting rod, then to the crankshaft.

Piston boss - The built-up area around the piston pin hole.

Piston collapse - The reduction in diameter of the piston skirt caused by heat and constant impact stresses.

Piston crown - The portion of the piston above the piston rings.

Piston head - See piston crown.

Piston lands - The portion of the piston between the ring grooves.

Piston pin (or wrist pin) - The cylindrical and usually hollow steel pin that passes through the piston. The piston pin fastens the piston to the upper end of the connecting rod.

Piston ring - The split ring fitted to the groove in a piston. The ring contacts the sides of the ring groove and also rubs against the cylinder wall, thus sealing space between piston and wall. There are two types of rings: Compression rings seal the compression pressure in the combustion chamber; oil rings scrape excessive oil off the cylinder wall.

Piston ring groove - The slots or grooves cut in piston head to hold piston rings in position.

Piston ring groove cleaning tool - A tool that cleans the carbon out of dirty ring grooves.

Piston ring side clearance - The space between the sides of the ring and the ring lands.

Piston skirt - The portion of the piston below the rings and the piston pin hole.

Piston slap - A sound made by a piston with excess skirt clearance as the crankshaft goes across top center.

Plastigage - A thin strip of plastic thread, available in different sizes, used for measuring clearances. For example, a strip of plastigage is laid across a bearing journal and mashed as parts are assembled. Then parts are disassembled and the width of the strip is measured to determine clearance between journal and bearing. Commonly used to measure crankshaft main-bearing and connecting rod bearing clearances.

Pop-up - A raised portion of a piston head.

Porting - A term which refers to smoothing out, aligning and/or enlarging intake passageway to valves.

Power stroke (firing stroke) - The portion of the piston's movement devoted to transmitting the power of the burning fuel mixture to the crankshaft. The power stroke occurs between top dead center and bottom dead center immediately following the compression stroke, during which both valves are closed and the air-fuel mixture burns, expands and forces the piston down to transmit power to the crankshaft.

Preignition - The ignition of the air-fuel mixture in the combustion chamber by some unwanted means, before the ignition spark occurs at the spark plug.

Preload - The amount of load placed on a bearing before actual operating loads are imposed. Proper preloading requires bearing adjustment and ensures alignment and minimum looseness in the system.

Press-fit - A tight fit between two parts that requires pressure to force the parts together. Also referred to as drive, or force, fit.

Prick punch - A small, sharp punch used to make punch marks on a metal surface.

Prussian blue - A blue pigment; in solution, useful in determining the area of contact between two surfaces. Prussian blue is commonly used to determine the width and location of the contact area between the valve face and the valve seat.

Puller - A special tool designed to remove a bearing, bushing, hub, sleeve, etc. There are many, many types of pullers.

Pushrod - The rod that connects the valve lifter to the rocker arm. The pushrod transmits cam lobe lift.

R

Race (bearing) - The inner or outer ring that provides a contact surface for balls or rollers in bearing.

Ramp - A gradual slope or incline on a cam to take up lash clearance.

Ream - To size, enlarge or smooth a hole by using a round cutting tool with fluted edges.

Rear main bearing seal - The large seal at the rear of the crankshaft that prevents oil from leaking into the bellhousing; there are two types, the rope seal and the rubber lip seal. Lip seals may be one-piece or two-piece.

Reciprocate - Move back and forth.

Retainer - A washer-like device that locks the keepers onto the valve stem and preloads the valve spring.

Ridge - See "Ring ridge."

Ring - See "Piston ring."

Ring gap - The distance between the ends of the piston ring when installed in the cylinder.

Ring job - The process of reconditioning the cylinders and installing new rings.

Ring ridge - The ridge, formed at the top of the cylinder above the upper limit of ring travel, as the cylinder wall below is worn away. In a worn cylinder, this area is of smaller diameter than the remainder of the cylinder, and will leave a ridge or ledge that must be removed.

Ring side clearance - The clearance between the top or bottom of a piston ring and the roof or floor, respectively, of its corresponding ring groove.

Rocker arm - A lever arm that rocks on a shaft or pivots on a stud as the cam moves the pushrod. The rocker arm converts the upward movement of the pushrod into a downward movement to open a valve.

Rocker arm clips - Clips which fit over the pushrod end of the rocker arm to prevent oil from being thrown off when valves are adjusted at their operating temperature.

Rocker arm pivot - The nut or fulcrum upon which the rocker arm pivots back and forth in a see-saw motion as it's alternately rocked one way by pushrod, then rocked other way by valve as it closes

Rod - See "Connecting rod."

Rod bearing - The bearing in the connecting rod in which a crankpin of the crankshaft rotates. Also called a connecting rod bearing.

Roller rocker arm - A special high-performance rocker arm that pivots on a roller bearing and has a roller tip.

Roller tappets or lifters - Valve lifters that have a roller in the end that contacts the camshaft. This reduces friction between the lobe and lifter. Used when special camshafts and high-tension springs have been installed.

Runout - Wobble. The amount a shaft rotates out-of-true.

S

Saddle - The upper main bearing seat.

Scored - Scratched or grooved, as a cylinder wall may be scored by abrasive particles moved up and down by the piston rings.

Haynes Chevrolet engine overhaul manual

Scraper ring - On a piston, an oil-control ring designed to scrape excess oil back down the cylinder and into the crankcase.

Scuffing - A type of wear in which there's a transfer of material between parts moving against each other; shows up as pits or grooves in the mating surfaces.

Sealant (gasket) - A thick, tacky compound, usually spread with a brush, which may be used as a gasket or sealant, to seal small openings or surface irregularities.

Seat - The surface upon which another part rests or seats. For example, the valve seat is the matched surface upon which the valve face rests. Also used to refer to wearing into a good fit; for example, piston rings seat after a few miles of driving.

Shimming - Placing a shim or spacer under weak valve springs or springs with a short free height; the shim, which is placed between the spring and the cylinder head, compresses the spring a little more to restore the spring's installed and open loads.

Shim-type head gaskets - A hard, thin, high-performance steel head gasket that raises the compression ratio.

Short block - An engine block complete with crankshaft and piston and, usually, camshaft assemblies.

Side clearance - The clearance between the sides of moving parts that don't serve as load-carrying surfaces.

Silent chain - A special quiet timing chain.

Slide hammer - A special puller that screws into or hooks onto the back of the a bearing; a heavy sliding handle on the shaft bottoms against the end of the shaft to knock the bearing free.

Slinger - A ring on a shaft that throws oil from the shaft before it gets to the oil seal.

Slipper skirt - A piston with a lower surface on the thrust surfaces only.

Sludge - An accumulation of water, dirt and oil in the oil pan; sludge is very viscous and tends to reduce lubrication.

Solid height - The height of a coil spring when it's totally compressed to the point at which each coil touches the adjacent coil. See "Coil binding."

Spacer - Another name for a valve spring shim. See "Shimming."

Spark knock - See "Detonation."

Spark tester - A device which indicates whether there's spark at the end of each plug wire. Used for quick check of the ignition system.

Split-lip-type rear main seal bearing - A two-piece neoprene seal; easier to install and has less friction than a rope-type rear main seal.

Spot-faced - On a connecting rod, a bolt-head seating surface that's machined so it describes a radius on the inboard-side of the bolt head as viewed from the top of the bolt head.

Spurt or squirt hole - A small hole in the connecting rod big end that indexes (aligns) with the oil hole in the crank journal. When the holes index, oil spurts out to lubricate the cylinder walls.

Static balance - The balance of an object while it's stationary.

Step - The wear on the lower portion of a ring land caused by excessive side and back-clearance. The height of the step indicates the ring's extra side clearance and the length of the step projecting from the back wall of the groove represents the ring's back clearance.

Stoichiometric - An air/fuel mixture that is balanced for greatest efficiency.

Stroboscope - See "Timing light."

Stroke - The distance the piston moves when traveling from top dead center to bottom dead center, or from bottom dead center to top dead center.

Stroked crankshaft - A crankshaft, either a special new one or a stock crank that's been reworked, which has the connecting rod throws offset so that the length of their stroke is increased.

Stud - A metal rod with threads on both ends.

Stud puller - A tool used to remove or install studs.

Sump - The lowest part of the oil pan. The part of oil pan that contains oil.

Swept volume - The volume displaced, or swept, by the piston as it travels from bottom dead center to top dead center.

T

Tang - A lip on the end of a plain bearing used to align the bearing during assembly.

Tap - To cut threads in a hole. Also refers to the fluted tool used to cut threads.

Tap and die set - Set of taps and dies for internal and external threading - usually covers a range of the most popular sizes.

Taper - A gradual reduction in the width of a shaft or hole; in an engine cylinder, taper usually takes the form of uneven wear, more pronounced at the top than at the bottom.

Tapered roller bearing - A bearing utilizing a series of tapered, hardened steel rollers operating between an outer and inner hardened steel race.

Tappet - See "Lifter."

TDC - Top Dead Center.

Threaded insert - A threaded coil that's used to restore the original thread size to a hole with damaged threads; the hole is drilled oversize and tapped, and the insert is threaded into the tapped hole.

Throws - The offset portions of the crankshaft to which the connecting rods are affixed.

Thrust bearing - The main bearing that has thrust faces to prevent excessive endplay, or forward and backward movement of the crankshaft.

Thrust plate - The small plate between the cam sprocket and the block, bolted to the front of the block.

Thrust washer - A bronze or hardened steel washer placed between two moving parts. The washer prevents longitudinal movement and provides a bearing surface for thrust surfaces of parts.

Timing - Delivery of the ignition spark or operation of the valves (in relation to the piston position) for the power stroke.

Timing chain - The chain, driven by a sprocket on the crankshaft, that drives the sprocket on the camshaft.

Timing gear - A gear on the crankshaft that drives the camshaft by meshing with a gear on its end.

Timing light - A stroboscopic light that's hooked up to the secondary ignition circuit to produce flashes of light in unison with the firing of a specific spark plug, usually the plug for the number one cylinder. When these flashes of light are directed on the whirling timing marks, the marks appear to stand still. By adjusting the distributor position, the timing marks can be properly aligned, and the timing is set.

Timing marks (ignition) - Marks, usually located on the vibration damper, used to synchronize the ignition system so the spark plugs will fire at the correct time.

Timing marks (valves) - One tooth on either the camshaft or crankshaft gear will be marked with an indentation or some other mark. Another mark will be found on the other gear between two of the teeth. The two gears must be meshed so that the marked tooth meshes with the marked spot on the other gear.

Tip - See "Valve stem tip."

Toe - The highest point on the cam lobe; the part of the lobe that raises the lifter to its highest point. Also called the nose.

Top dead center - The piston position when the piston has reached the upper limit of its travel in the cylinder and the center line of the connecting rod is parallel to the cylinder walls.

Tolerance - The amount of variation permitted from an exact size of measurement. Actual amount from smallest acceptable dimension to largest acceptable dimension.

Torque - A turning or twisting force, such as the force imparted on a fastener by a torque wrench. Usually expressed in foot-pounds (ft-lbs).

Torque plate - A stout steel plate with four large-diameter holes centered on the engine bores to allow clearance for boring and honing the cylinders, bolted onto the block as a temporary "cylinder head" during machining. Using a torque plate prevents bore distortion while the cylinders are being bored and honed.

Torsional vibration - The rotary motion that causes a twist-untwist action on a vibrating shaft, so that a part of the shaft repeatedly moves ahead of, or lags behind, the remainder of the shaft; for example, the action of a crankshaft responding to the cylinder firing impulses.

Twist drill - A metal cutting drill with spiral flutes (grooves) to permit exit of chips while cutting.

U

Umbrella - An oil deflector placed near the valve tip to throw oil from the valve stem area.

Undercut - A machined groove below the normal surface.

Undersize bearings - Smaller diameter bearings used with re-ground crankshaft journals.

V

Valve - A device used to either open or close an opening. Usually used to allow or stop the flow of a liquid or a gas. There are many different types.

Valve clearance - The clearance between the valve tip (the end of the valve stem) and the rocker arm. The valve clearance is measured when the valve is closed.

Valve duration - The length of time, measured in degrees of engine crankshaft rotation, that the valve remains open.

Valve face - The outer lower edge of the valve. The valve face contacts the valve seat in the cylinder head when the valve is closed.

Valve float - The condition which occurs when the valves are forced back open before they've had a chance to seat. Valve float is usually caused by extremely high rpm.

Valve grinder - A special electrically-powered grinding machine designed to remove old metal from the valve face.

Valve grinding - Refacing a valve in a valve-refacing machine.

Valve guide - The cast-iron bore that's part of the head, or the bronze or silicon-bronze tube that's pressed into the head, to provide support and lubrication for the valve stem.

Valve keeper - Also referred to as valve key. Small half-cylinder of steel that snaps into a groove in the upper end of valve stem. Two keepers per valve are used. Designed to secure valve spring, valve retainer and valve stem together.

Valve lift - The distance a valve moves from its fully closed to its fully open position.

Valve lifter - A cylindrical device that contacts the end of the cam lobe and the lower end of the pushrod. The lifter rides on the camshaft. When the cam lobe moves it upward, it pushes on the pushrod, which pushes on the lifer and opens the valve. Referred to as a lifter, tappet, valve tappet or cam follower.

Valve lock - See "Valve keeper."

Valve margin - The thickness of the valve head at its outside diameter, between the top of the valve head and the outer edge of the face.

Valve overlap - The number of degrees of crankshaft rotation during which both the intake and the exhaust valve are partially open (the intake is starting to open while the exhaust is not yet closed).

Valve port - An opening, through the cylinder head, from the intake or exhaust manifold to the valve seat.

Valve rotator - The device placed on the end of the valve stem to promote longer valve life. When the valve is opened and closed, the valve will rotate a small amount with each opening and closing.

Valve seat - The surface against which a valve comes to rest to provide a seal against leaking.

Valve seat insert - The hardened, precision ground metal ring pressed into the combustion chamber to provide a sealing surface for the valve face (usually for the exhaust valves) and transfer heat into the head. Inserts are made of special metals able to withstand very high temperatures.

Valve spring - A coil spring designed to close the valve against its seat.

Valve spring squareness - How straight a spring stands on a flat surface, or how much it tilts. The more "square" a spring is, the more evenly it loads the retainer around its full circumference. Uneven retainer loading increases stem and guide wear.

Valve stem - The long, thin, cylindrical bearing surface of the valve that slides up and down in the valve guide.

Valve stem skeal - A device placed on or around the valve stem to reduce the amount of oil that can get on the stem and then work its way down into the combustion chamber.

Valve stem tip - The upper end of the valve stem; the tip is the contact point with the rocker arm.

Valve tappet - See "Valve lifter."

Valve timing - The timing of the opening and closing of the valves in relation to the piston position.

Valve tip - The upper end of the valve that contacts the rocker arm.

Valve train - The valve-operating mechanism of an engine; includes all components from the camshaft to the valve.

Valve umbrella - A washer-like unit that's placed over the end of the valve stem to prevent the entry of excess oil between the stem and the guide.

Varnish - The deposits on the interior of the engine caused by engine oil breaking down under prolonged heat and use. Certain constituents of oil deposit themselves in hard coatings of varnish.

Vernier caliper - A precision measuring instrument that measures inside and outside dimensions. Not quite as accurate as a micrometer, but more convenient.

Vibration damper - A cylindrical weight attached to the front of the crankshaft to minimize torsional vibration (the twist-untwist actions of the crankshaft caused by the cylinder firing impulses). Also called a harmonic balancer.

Volumetric efficiency - A comparison between actual volume of fuel mixture drawn in on intake stroke and what would be drawn in if cylinder were to be completely filled.

W

Water jacket - The spaces around the cylinders, between the inner and outer shells of the cylinder block or head, through which coolant circulates.

Web - A supporting structure across a cavity.

Wet manifold - An intake manifold that carries coolant through integral passages.

Wet sleeve - A cylinder sleeve whose outside surface is in direct contact with coolant.

Woodruff key - A key with a radiused backside (viewed from the side).

Wrist pin - See "Piston pin."

Notes

Appendix

A Booster battery (jump) starting

Observe the following precautions when using a booster battery to start a vehicle:

a) *Before connecting the booster battery, make sure the ignition switch is in the Off position.*
b) *Turn off the lights, heater and other electrical loads.*
c) *Your eyes should be shielded. Safety goggles are a good idea.*
d) *Make sure the booster battery is the same voltage as the dead one in the vehicle.*
e) *The two vehicles MUST NOT TOUCH each other.*
f) *Make sure the transmission is in Neutral (manual transaxle) or Park (automatic transaxle).*
g) *If the booster battery is not a maintenance-free type, remove the vent caps and lay a cloth over the vent holes.*

Connect the red jumper cable to the positive (+) terminals of each battery.

Connect one end of the black cable to the negative (-) terminal of the booster battery. The other end of this cable should be connected to a good ground on the engine block **(see illustration)**. Make sure the cable will not come into contact with the fan, drivebelts or other moving parts of the engine.

Start the engine using the booster battery, then, with the engine running at idle speed, disconnect the jumper cables in the reverse order of connection.

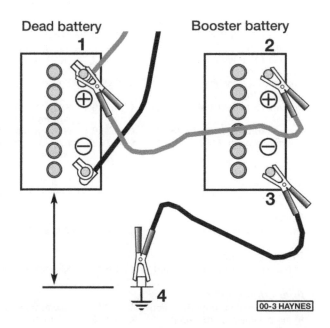

Make the booster battery cable connections in the numerical order shown (note that the negative cable of the booster battery is NOT attached to the negative terminal of the dead battery)

B Conversion factors

Length (distance)

Inches (in)	X	25.4	= Millimetres (mm)	X 0.0394	= Inches (in)
Feet (ft)	X	0.305	= Metres (m)	X 3.281	= Feet (ft)
Miles	X	1.609	= Kilometres (km)	X 0.621	= Miles

Volume (capacity)

Cubic inches (cu in; in^3)	X	16.387	= Cubic centimetres (cc; cm^3)	X 0.061	= Cubic inches (cu in; in^3)
Imperial pints (Imp pt)	X	0.568	= Litres (l)	X 1.76	= Imperial pints (Imp pt)
Imperial quarts (Imp qt)	X	1.137	= Litres (l)	X 0.88	= Imperial quarts (Imp qt)
Imperial quarts (Imp qt)	X	1.201	= US quarts (US qt)	X 0.833	= Imperial quarts (Imp qt)
US quarts (US qt)	X	0.946	= Litres (l)	X 1.057	= US quarts (US qt)
Imperial gallons (Imp gal)	X	4.546	= Litres (l)	X 0.22	= Imperial gallons (Imp gal)
Imperial gallons (Imp gal)	X	1.201	= US gallons (US gal)	X 0.833	= Imperial gallons (Imp gal)
US gallons (US gal)	X	3.785	= Litres (l)	X 0.264	= US gallons (US gal)

Mass (weight)

Ounces (oz)	X	28.35	= Grams (g)	X 0.035	= Ounces (oz)
Pounds (lb)	X	0.454	= Kilograms (kg)	X 2.205	= Pounds (lb)

Force

Ounces-force (ozf; oz)	X	0.278	= Newtons (N)	X 3.6	= Ounces-force (ozf; oz)
Pounds-force (lbf; lb)	X	4.448	= Newtons (N)	X 0.225	= Pounds-force (lbf; lb)
Newtons (N)	X	0.1	= Kilograms-force (kgf; kg)	X 9.81	= Newtons (N)

Pressure

Pounds-force per square inch (psi; lbf/in^2; lb/in^2)	X	0.070	= Kilograms-force per square centimetre (kgf/cm^2; kg/cm^2)	X 14.223	= Pounds-force per square inch (psi; lbf/in^2; lb/in^2)
Pounds-force per square inch (psi; lbf/in^2; lb/in^2)	X	0.068	= Atmospheres (atm)	X 14.696	= Pounds-force per square inch (psi; lbf/in^2; lb/in^2)
Pounds-force per square inch (psi; lbf/in^2; lb/in^2)	X	0.069	= Bars	X 14.5	= Pounds-force per square inch (psi; lbf/in^2; lb/in^2)
Pounds-force per square inch (psi; lbf/in^2; lb/in^2)	X	6.895	= Kilopascals (kPa)	X 0.145	= Pounds-force per square inch (psi; lbf/in^2; lb/in^2)
Kilopascals (kPa)	X	0.01	= Kilograms-force per square centimetre (kgf/cm^2; kg/cm^2)	X 98.1	= Kilopascals (kPa)

Torque (moment of force)

Pounds-force inches (lbf in; lb in)	X	1.152	= Kilograms-force centimetre (kgf cm; kg cm)	X 0.868	= Pounds-force inches (lbf in; lb in)
Pounds-force inches (lbf in; lb in)	X	0.113	= Newton metres (Nm)	X 8.85	= Pounds-force inches (lbf in; lb in)
Pounds-force inches (lbf in; lb in)	X	0.083	= Pounds-force feet (lbf ft; lb ft)	X 12	= Pounds-force inches (lbf in; lb in)
Pounds-force feet (lbf ft; lb ft)	X	0.138	= Kilograms-force metres (kgf m; kg m)	X 7.233	= Pounds-force feet (lbf ft; lb ft)
Pounds-force feet (lbf ft; lb ft)	X	1.356	= Newton metres (Nm)	X 0.738	= Pounds-force feet (lbf ft; lb ft)
Newton metres (Nm)	X	0.102	= Kilograms-force metres (kgf m; kg m)	X 9.804	= Newton metres (Nm)

Vacuum

Inches mercury (in. Hg)	X	3.377	= Kilopascals (kPa)	X 0.2961	= Inches mercury
Inches mercury (in. Hg)	X	25.4	= Millimeters mercury (mm Hg)	X 0.0394	= Inches mercury

Power

Horsepower (hp)	X	745.7	= Watts (W)	X 0.0013	= Horsepower (hp)

Velocity (speed)

Miles per hour (miles/hr; mph)	X	1.609	= Kilometres per hour (km/hr; kph)	X 0.621	= Miles per hour (miles/hr; mph)

Fuel consumption*

Miles per gallon, Imperial (mpg)	X	0.354	= Kilometres per litre (km/l)	X 2.825	= Miles per gallon, Imperial (mpg)
Miles per gallon, US (mpg)	X	0.425	= Kilometres per litre (km/l)	X 2.352	= Miles per gallon, US (mpg)

Temperature

Degrees Fahrenheit = (°C x 1.8) + 32   Degrees Celsius (Degrees Centigrade; °C) = (°F - 32) x 0.56

*It is common practice to convert from miles per gallon (mpg) to litres/100 kilometres (l/100km), where mpg (Imperial) x l/100 km = 282 and mpg (US) x l/100 km = 235

C Electrical glossary

A

Active material - The material on the negative and positive battery plates that interacts with the electrolyte to produce a charge.

AC generator - An electromechanical device that generates alternating current (AC), commonly known as an alternator. Usually belt-driven off the engine. Provides maximum output at relatively low rpm. Used on all modern vehicles. Requires a rectifier to convert AC to direct current (DC), which is used by automotive electrical system.

Aftermarket parts - Components that can be added to a vehicle after its manufacture. These parts are often accessories and should not be confused with original equipment manufacturer (OEM) service or replacement parts.

Alligator clip - A long-nosed spring-loaded metal clip with meshing teeth. Used to make temporary electrical connections.

Alloy - A mixture of two or more materials.

Alternating current (AC) - An electric current, generated through magnetism, whose polarity constantly changes back and forth from positive to negative.

Alternator - A device used in automobiles to produce electric current. The alternator's AC output is rectified to direct current before it reaches the vehicle's electrical system.

Ammeter - 1. An instrument for measuring current flow. An ammeter can be designed to measure alternating or direct current. 2. An instrument panel gauge used to measure the flow rate of current into or out of the battery. The ammeter is calibrated in amperes for both charge and discharge rates, in ranges of 20, 30 or more amperes.

Amperage - The total amount of current (amperes) flowing in a circuit.

Ampere (amp) - The unit of measurement for the flow of electrons in a circuit. The amount of current produced by one volt acting through a resistance of one ohm (1 coulomb per second).

Ampere hour - A unit of measurement for battery capacity, obtained by multiplying the current (in amperes) by the time (in hours) during which the current is delivered.

Analog gauge - see "gauge."

Analog signal - A signal which varies in exact proportion to a measured quantity, such as pressure, temperature, speed, etc.

Arcing - When electricity leaps the gap between two electrodes, it is said to be "arcing."

Armature - The rotating part of a generator or motor. Actually a coil of wires wrapped in a specific pattern, which rotates on a shaft.

Atoms - The small particles which make up all matter. Atoms are made up of a positive-charged nucleus with negative-charged electrons whirling around in orbits.

B

Battery - A group of two or more cells connected together for the production of an electric current by converting chemical energy into electrical energy. A battery has two poles - positive and negative. The amount of positive and negative energy is called potential.

Battery charging - The process of energizing a battery by passing electric current through the battery in a reverse direction.

Battery ratings - Performance standards conducted under laboratory conditions to describe a battery's reserve capacity and cold-crank capacities. The amp-hour rating is no longer in widespread use. See "cold crank rating."

Battery state of charge - The available amount of energy in a battery in relation to that which would ordinarily available be if the battery was fully charged.

Battery voltage - A figure determined by the number of cells in a battery. Because each cell generates about two volts, a six cell battery has 12 volts.

Bendix inertia drive - A self-engaging and releasing starter drive mechanism. The pinion gear moves into engagement when the starter motor spins and disengages when the engine starts.

Bound electron - An electron whose orbit is near the nucleus of an atom and is strongly attracted to it.

Brush - A spring-loaded block of carbon or copper that rubs against a commutator or slip ring to conduct current. A key component in all alternators and starters.

Bulkhead connector - An OEM device used to connect wiring inside the vehicle body with wiring outside the body. Usually located at the bulkhead or firewall.

Butt connector - A solderless connector used to permanently join two wire ends together.

Bypass - A shunt, or parallel path, around one or more elements in a circuit.

C

Cable - An assembly of one or more conductors, usually individually insulated and enclosed in a protective sheath.

Capacity - The current output capability of a cell or battery, usually expressed in ampere hours.

Cell - In a storage battery, one of the sets of positive and negative plates which react with electrolyte to produce an electric current.

Charge - A definite quantity of electricity.

Charge (recharge) - To restore the active materials in a battery cell by electrically reversing the chemical action.

Circuit - An electrical path - from the source (battery or generator) through the load (such as a light bulb) and back to the source - through which current flows. A typical circuit consists of a battery, wire, load (light or motor) and switch. See "simple circuit" and "single-wire circuit."

Circuit breaker - A circuit-protection device that automatically opens or breaks an overloaded circuit. The typical circuit breaker usually consists of movable points that open if the preset ampere load is exceeded. Some circuit breakers are self-resetting; others require manual resetting.

Closed circuit - A circuit which is uninterrupted from the current source, through the load and back to the current source.

Closed-end connector - Solderless connector shaped like a hat. Used to join two, three or more wires together. Similar to wire connectors used in home wiring, but installed by crimping instead of twisting.

Clutch interlock switch - A switch that prevents the vehicle from starting unless the clutch is pressed.

Coil - Any electrical device or component consisting of wire loops wrapped around a central core. Coils depend on one of two electrical properties for operation, depending on the application (electromagnetism or induction).

Cold-crank rating - The minimum number of amperes a fully charged 12-volt battery can deliver for 30 seconds at 0-degrees F without falling below 7.2 battery volts.

Commutator - A series of copper bars insulated from each other and connected to the armature windings at the end of the armature. Provides contact with fixed brushes to draw current from (generator) or bring current to (starter) the armature.

Conductance - A measure of the ease with which a conductor allows electron flow. In DC circuits, conductance is the reciprocal of resistance.

Conduction - The transmission of heat or electricity through, or by means of, a conductor.

Conductor - Any material - usually a wire or other metallic object - made up of atoms whose free electrons are easily dislodged, permitting easy flow of electrons from atom to atom. Examples are copper, aluminum and steel. Conductors are all metals. The metal part of an insulated wire is often called the conductor.

Constant voltage regulator (CVR) - A device used to maintain a constant voltage level in a circuit, despite fluctuations in system voltage. CVRs are wired into some gauge circuits so voltage fluctuations won't affect accuracy of the gauge readings.

Contact - One of the contact-carrying parts of a relay or switch that engages or disengages to open or close the associated electrical circuits.

Continuity - A continuous path for the flow of an electrical current.

Conventional theory - In this theory, the direction of current flow was arbitrarily chosen to be from the positive terminal of the voltage source, through the external circuit, then back to the negative terminal of the source.

Coulomb - The unit of quantity of electricity or charge. The electrons that pass a given point in one second when the current is maintained at one ampere. Equal to an electrical charge of 6.25×10^{18} electrons passing a point in one second. See "ampere."

Current - The movement of free electrons along a conductor. In automotive electrical work, electron flow is considered to be from positive to negative. Current flow is measured in amperes.

Cycle - A recurring series of events which take place in a definite order.

D

DC generator - An electromechanical device that generates direct current. Usually belt-driven off the engine. Because the DC generator requires high rpm for maximum output, it's no longer used in production automobiles.

Deep cycling - The process of discharging a battery almost completely before recharging.

Digital gauge - See "gauge."

Diode - A semiconductor which permits current to flow in only one direction. Diodes are used to rectify current from AC to DC.

Direct current (DC) - An electrical current which flows steadily in only one direction.

Discharge - Generally, to draw electric current from the battery. Specifically, to remove more energy from a battery than is being replaced. A discharged battery is of no use until it's recharged.

Disconnect terminal - Solderless connectors in male and female forms, intended to be easily disconnected and connected. Typically, a blade or pin (male connector) fits into a matching receptacle or socket (female connector). Many components have built-in (blade) terminals that require a specialized female connector.

Display - Any device that conveys information. In a vehicle, displays are either lights, gauges or buzzers. Gauges may be analog or digital.

DPDT - A double-pole, double-throw switch.

DPST - A double-pole, single-throw switch.

Draw - The electric current required to operate an electrical device.

Drive - A device located on the starter to allow for a method of engaging the starter to the flywheel.

E

Electric - A word used to describe anything having to do with electricity in any form. Used interchangeably with electrical.

Electrical balance - An atom or an object in which positive and negative charges are equal.

Electricity - The movement of electrons from one body of matter to another.

Electrochemical - The production of electricity from chemical reactions, as in a battery.

Electrolyte - A solution of sulfuric acid and water used in the cells of a battery to activate the chemical process which results in an electrical potential.

Electromagnet - A soft-iron core which is magnetized when an electric current is passed through a coil of wire surrounding it.

Electromagnetic - Having both electrical and magnetic properties.

Electromagnetism - The magnetic field around a conductor when a current is flowing through the conductor.

Electromechanical - Any device which uses electrical energy to produce mechanical movement.

Electrons - Those parts of an atom which are negatively charged and orbit around the nucleus of the atom.

Electron flow - The movement of electrons from a negative to a positive point on a conductor, or through a liquid, gas or vacuum.

Electron theory - States that all matter is made up of atoms which are made up of a nucleus and electrons. Free electrons moving from one atom to another, in a single direction, produce what is known as electricity.

Electronics - The science and engineering concerned with the behavior of electrons in devices and the utilization of such devices. Especially devices utilizing electron tubes or semiconductor devices.

Energy - The capacity for performing work.

EVR - Electronic Voltage Regulator; a type of regulator that uses all solid state devices to perform the regulatory functions.

F

Field - An area covered or filled with a magnetic force. Common terminology for field magnet, field winding, magnetic field, etc.

Field coil - A coil of insulated wire, wrapped around an iron or steel core, that creates a magnetic field when current is passed through the wire.

Filament - A resistance in an electric light bulb which glows and produces light when an adequate current is sent through it.

Fluorescent - Having the property of giving off light when bombarded by electrons or radiant energy.

Flux - The lines of magnetic force flowing in a magnetic field.

Flywheel - A large wheel attached to the crankshaft at the rear of the engine.

Flywheel ring gear - A large gear pressed onto the circumference of the flywheel. When the starter gear engages the ring gear, the starter cranks the engine.

Free electron - An electron in the outer orbit of an atom, not strongly attracted to the nucleus; it can be easily forced out of its orbit.

Fuse - A circuit-protection device containing a soft piece of metal which is calibrated to melt at a predetermined amp level and break the circuit.

Fuse block - An insulating base on which fuse clips or other contacts are mounted.

Fuse link - See "fusible link."

Fuse panel - A plastic or fiberboard assembly that permits mounting several fuses in one centralized location. Some fuse panels are part of, or contain, a terminal block (see "terminal block").

Fuse wire - A wire made of an alloy which melts at a low temperature.

Fusible link - A circuit protection device consisting of a conductor surrounded by heat-resistant insulation. The conductor is two gages smaller than the wire it protects, so it acts as the weakest link in the circuit. Unlike a blown fuse, a failed fusible link must be cut from the wire for replacement.

G

Gage - A standard SAE designation of wire sizes, expressed in AWG (American Wire Gage). The larger the gage number, the smaller the wire. Metric wire sizes are expressed in cross-sectional area, which is expressed in square millimeters. Sometimes the spelling "gauge" is also used to designate wire size. Using this spelling, however, avoids confusion with instrument panel displays (see "gauge").

Gassing - The breakdown of water into hydrogen and oxygen gas in a battery.

Gauge - An instrument panel display used to monitor engine conditions. A gauge with a movable pointer on a dial or a fixed scale is an analog gauge. A gauge with a numerical readout is called a digital gauge. Also refers to measuring device used to check regulator point openings.

Generator - An engine-driven device that produces an electric current through magnetism by converting rotary motion into electrical potential (see "AC generator" and "DC generator").

Grid - A lead screen that is pasted with active materials to form a negative or positive battery plate.

Grommet - A donut shaped rubber or plastic part used to protect wiring that passes through a panel, firewall or bulkhead.

Ground - The connection made in grounding a circuit. In a single-wire system, any metal part of the car's structure that's directly or indirectly attached to the battery's negative post. Used to conduct current from a load back to the battery. Self-grounded components are attached directly to a grounded metal part through their mounting screws. Components mounted to nongrounded parts of a vehicle require a ground wire attached to a known good ground.

H

Halogen light - A special bulb that produces a brilliant white light. Because of its high intensity, a halogen light is often used for fog lights and driving lights.

Harness - A bundle of electrical wires. For convenience in handling and for neatness, all wires going to a certain part of the vehicle are bundled together into a harness.

Harness ties - Self-tightening nylon straps used to bundle wires into harnesses. Available in stock lengths that can be cut to size after installation. Once tightened, they can't be removed unless they're cut from the harness.

Harness wrap - One of several materials used to bundle wires into manageable harnesses. See "loom," "split loom," "loom tape" and "harness ties").

Hot - Connected to the battery positive terminal, energized.

Hydrogen gas - The lightest and most explosive of all gases. Emitted from a battery during charging procedures. This gas is very dangerous and certain safety precautions must be observed.

Hydrometer - A syringe-like instrument used to measure the specific gravity of a battery's electrolyte.

I

IAR - Integral Alternator/Regulator; a type of regulator mounted at the rear of the alternator.

Ignition switch - A key-operated switch that opens and closes the circuit that supplies power to the ignition and electrical system.

Indicator light - An instrument-panel display used to convey information or condition of the monitored circuit or system. See "warning light."

Induced - Produced by the influence of a magnetic or electrical field.

Induced current - The current generated in a conductor as it moves through a magnetic field, or as a magnetic field is moved across a conductor.

Induced voltage - The voltage produced as a result of an induced current flow.

Inductance - That property of a coil or other electrical device which opposes any change in the existing current, present only when an alternating or pulsating current is flowing. Has no effect on the flow of direct, or static, current.

Induction - The process by which an electrical conductor becomes charged when near another charged body. Induction occurs through the influence of the magnetic fields surrounding the conductors.

Input - 1. The driving force applied to a circuit or device. 2. The terminals (or other connection) where the driving force may be applied to a circuit or device.

Instrument Voltage Regulator - See "Constant Voltage Regulator."

Insulator - A material that has few or no free electrons that readily leave their orbits. A non-conducting material used for insulating wires in electrical circuits. Cloth, glass, plastic and rubber are typical examples. Wires for modern vehicles use plastic insulation.

Integral - Formed as a unit with another part.

Intercell connector - A lead strap or connector that connects the cells in a battery.

Intermittent - Coming or going at intervals; not continuous.

Ion - An atom having an imbalanced charge due to the loss of an electron or proton. An ion may be either positively charged (have a deficiency of electrons) or negatively charged (have a surplus of electrons).

J

Jumper - A short length of wire used as a temporary connection between two points.

Junction - 1. A connection between two or more components, conductors or sections of transmission line. 2. Any point from which three or more wires branch out in a circuit.

Junction box - A box in which connections are made between different cables.

L

Lead-acid battery - A common automotive battery in which the active materials are lead, lead peroxide and a solution of sulfuric acid.

Lead dioxide - A combination of lead and oxygen, as found in the storage battery. Lead dioxide is reddish brown in color.

Lead sulfate - A combination of lead, oxygen and sulfur, as found in the storage battery.

Light - An electrical load designed to emit light when current flows through it. A light consists of a glass bulb enclosing a filament and a base containing the electrical contacts. Some lights, such as sealed beam headlights, also contain a built-in reflector.

Load - Any device that uses electrical current to perform work in a vehicle electrical system. Lights and motors are typical examples.

Loom - A harness covering. Older vehicles used woven-cloth loom; most modern vehicles use a corrugated-plastic loom or split loom.

Loom tape - A nonadhesive tape used as a harness wrap. Adhesive-type tapes, including electrical tapes, are not recommended for wrapping harnesses. Often, a piece of shrink wrap is used at tape ends to keep the tape from unraveling.

M

Magnet - A material that attracts iron and steel. Temporary magnets are made by surrounding a soft-iron core with a strong electromagnetic field. Permanent magnets are made with steel.

Magnetic field - The field produced by a magnet or a magnetic influence. A magnetic field has force and direction.

Magnetic poles - The ends of a bar or horseshoe magnet.

Magnetism - A property of the molecules of certain materials, such as iron, that allows the substance to be magnetized.

Meter - An electrical or electronic measuring device.

Module - A combination of components packaged in a single unit with a common mounting and providing some complete function.

Motor - An electromagnetic apparatus used to convert electrical energy into mechanical energy.

Multimeter - A test instrument with the capability to measure voltage, current and resistance.

N

Negative charge - The condition when an element has more than a normal quantity of electrons.

Negative ion - An atom with more electrons than normal. A negative ion has a negative charge.

Negative terminal - The terminal on a battery which has an excess of electrons. A point from which electrons flow to the positive terminal.

Neutral - Neither positive nor negative, or in a natural condition. Having the normal number of electrons, i.e. the same number of electrons as protons.

Neutral start switch - On vehicles with an automatic transmission, a switch that prevents starting if the vehicle is not in Neutral or Park.

Neutron - A particle within the nucleus of an atom. A neutron is electrically neutral.

Nichrome - A metallic compound containing nickel and chromium, used in making high resistances.

North pole - The pole of a magnet from which the lines of force are emitted. The lines of force travel from the north to the south pole.

Nucleus - The core of an atom. The nucleus contains protons and neutrons.

Nylon ties - See "harness ties."

O

OEM (Original Equipment Manufacturer) - A designation used to describe the equipment and parts installed on a vehicle by the manufacturer, or those available from the vehicle manufacturer as replacement parts. See "aftermarket parts."

Ohm - The practical unit for measuring electrical resistance.

Ohmmeter - An instrument for measuring resistance. In automotive electrical work, it's often used to determine the resistance that various loads contribute to a circuit or system.

Ohm's Law - The electrical formula that describes how voltage, current and resistance are related. The basic formula is E (electrical pressure in volts) = I (current flow in amperes) X R (resistance in ohms). What does it mean? Simply put, amperage varies in direct ratio to voltage and in inverse ratio to resistance.

Open circuit - An electrical circuit that isn't complete because of a broken or disconnected wire.

Open circuit voltage - The battery voltage when the battery has no closed circuit across the posts and is not delivering or receiving voltage.

Orbit - The path followed by an electron around the nucleus.

Output - The current, voltage, power or driving force delivered by a device or circuit. The terminals or connections where the current can be measured.

Overrunning clutch - A device located on the starter to allow for a method of engaging the starter with the flywheel. The overrunning clutch uses a shift lever to actuate the drive pinion to provide for a positive meshing and de-meshing of the pinion with the flywheel ring gear.

P

Parallel circuit - A method or pattern of connecting units in an electrical circuit so that they're connected negative-to-negative and positive-to-positive. In a parallel circuit, current can flow independently through several components at the same time. See "series circuit."

Permanent magnet - A magnet made of tempered steel which holds its magnetism for a long period of time.

Pinion - A small gear that either drives or is driven by a larger gear.

Plate - A battery grid that's pasted with active materials and given a forming charge which results in a negative or positive charge. Plates are submerged, as elements, in the electrolyte and electricity is produced from the chemical reactions between the plates and the electrolyte.

Polarity - The quality or condition of a body which has two opposite properties or directions; having poles, as in an electric cell, a magnet or a magnetic field.

Polarity-protected connector - A multiple-cavity connector that can be connected in only one way, either to a mating connector or to a component.

Pole - A positive (or negative) terminal in a cell or battery; the ends of a magnet (north or south).

Pole shoe - The part of a starter that's used to hold the field coils in place in their proper positions; consists of a soft-iron core which is wound with heavy copper ribbons.

Positive - Designating or pertaining to a kind of electrical potential.

Positive terminal - The battery terminal to which current flows.

Post - A round, tapered battery terminal that serves as a connection for battery cables.

Potential - Latent, or unreleased, electrical energy.

Power supply - A unit that supplies electrical power to a unit. For example, a battery.

Printed circuit - An electrical conductor consisting of thin metal foil paths attached to a flexible plastic backing. Also called a PC board. PC boards are used primarily in OEM instrument clusters and other electronic devices.

Proton - A positively-charged particle in the nucleus of an atom.

Q

Quick charger - A battery charger designed to allow the charging of a battery in a short period of time.

R

Rectification - The process of changing alternating current to direct current.

Rectifier - A device in the electrical system used to convert alternating current to direct current.

Regulator - A device used to regulate the output of a generator or alternator by controlling the current and voltage.

Relay - An electromagnetic device that opens or closes the flow of current in a circuit.

Resistance - The resistance to electron flow present in an electrical circuit, expressed in ohms.

Resistor - Any conductor that permits electron movement but retards it. Tungsten and nickel are typical resistors.

Rheostat - A variable resistor, operated by a knob or handle, used to vary the resistance in a circuit. A rheostat consists of a coil of resistance wire and a movable contact or wiper that creates more or less resistance in the circuit, depending on how many coil windings it allows the current to pass through. The dimmer control for instrument panel illumination is an example.

Ring terminal - A conductor used to attach a wire to a screw or stud terminal. The ring is sized to the mating screw. Ring terminals are the connectors least likely to vibrate loose in rugged applications. Comes in soldered and unsoldered versions.

Rotor - That part of an alternator which rotates inside the stator.

S

Schematic - A drawing system for portraying the components and wires in a vehicle's electrical system, using standardized symbols.

Secondary - The output winding of a transformer, i.e. the winding in which current flow is due to inductive coupling with another coil called the primary.

Sending unit - Used to operate a gauge or indicator light. In indicator light circuits, contains contact points, like a switch. In gauge circuits, contains a variable resistance that modifies current flow in accordance with the condition or system being monitored.

Separator - A thin sheet of non-conducting material that is placed between the negative and positive plates in an element to prevent the plates from touching.

Series circuit - A circuit in which the units are consecutively connected or wired positive to negative and in which current has to pass through all components, one at a time.

Series/parallel circuit - A circuit in which some components are wired in series, others in parallel. An example: Two loads wired in parallel with each other, but in series with the switch that controls them.

Short circuit - An unintentional routing of a current, bypassing part of the original circuit.

Shorted winding - The winding of a field or armature that's grounded because of accidental or deliberate reasons.

Shrink wrap - An insulating material used to protect wire splices and junctions at terminals. Upon application of open flame or heat, the wrap shrinks to fit tightly on the wire or terminal.

Shunt - 1. Connected in parallel with some other part. 2. A precision low value resistor placed across the terminals of an ammeter to increase its range.

Simple circuit - The simplest circuit includes an electrical power source, a load and some wire to connect them together.

Single-wire circuit - Generally used in production vehicles, in which one wire brings current to the load and the vehicle's frame acts as the return path (ground).

Slip ring - A device for making electrical connections between stationary and rotating contacts.

Snap-splice connector - Solderless connector used to tap an additional wire into an existing wire without cutting the original. Often used in installing trailer wiring to a tow vehicle.

Solder - An alloy of lead and tin which melts at a low temperature and is used for making electrical connections.

Solderless connector - Any connector or terminal that can be installed to a wire without the use of solder. They're usually crimped in place using a special crimping tool. Ring terminals, spade terminals, disconnect terminals, butt connectors, closed-end connectors and snap-splice connectors are typical examples. Ring and spade terminals also come in soldered versions.

Solenoid - An electromechanical device consisting of a tubular coil surrounding a movable metal core, or plunger, which moves when the coil is energized. The movable core is connected to various mechanisms to accomplish work.

Spade terminal - A terminal used to connect a wire to a screw or stud terminal. The spade terminal has two forked ends, either straight or with upturned tips. They're more convenient to install than ring terminals, but slightly less secure for rugged applications. Comes in soldered and unsoldered versions.

SPDT - A single-pole, double-throw switch.

Specific gravity - The measure of a battery's charge, made by comparing the relative weight of a volume of its electrolyte to the weight of an equal volume of pure water, which has an assigned value of 1.0. A fully charged battery will have a specific gravity reading of 1.260. See "hydrometer."

Split loom - Flexible, corrugated conduit used to bundle wires into harnesses.

SPST - A single-pole, single-throw switch.

Starter - A device used to supply the required mechanical force to turn over the engine for starting.

Stator - In an alternator, the part which contains the conductors within which the field rotates.

Sulfuric acid - A heavy, corrosive, high-boiling liquid acid that is colorless when pure. Sulfuric acid is mixed with distilled water to form the electrolyte used in storage batteries.

Supply voltage - The voltage obtained from the power supply to operate a circuit.

Switch - An electrical control device used to turn a circuit on and off, or change the connections in the circuit. Switches are described by the number of poles and throws they have. See "SPST, SPDT, DPST and DPDT."

T

Tachometer - A device that measures the speed of an engine in rpm.

Terminal - A device attached to the end of a wire or to an apparatus for convenience in making electrical connections.

Terminal block - A plastic or resin assembly containing two rows of terminals screws. Used to join the circuits in several wiring harnesses.

Test light - A test instrument consisting of an indicator light wired into the handle of a metal probe. When the probe

contacts a live circuit, current flows through the light, lighting it, and to ground through an attached lead and alligator clip. Used to test for voltage in live circuits only.

Test light (self-powered) - A test device containing an indicator light and a built-in battery. Used to test continuity of circuits not containing voltage at the time of the test. Used to test continuity in a harness before it's installed in the vehicle. Also called a continuity tester.

Thermal relay - A relay actuated by the heating effect of the current flowing through it.

Thermistor - The electrical element in a temperature sending unit that varies its resistance in proportion to temperature. Unlike most electrical conductors, in which resistance increases as temperature rises, resistance in a thermistor decreases. Thermistors are made from the oxides of cobalt, copper, iron or nickel.

Tracer - A stripe of a second color applied to a wire insulator to distinguish that wire from another one with the same color insulator.

Transformer - An apparatus for transforming an electric current to a higher or lower voltage without changing the total energy.

V

Variable resistor - A wire-wound or composition resistor with a sliding contact for changing the resistance. See "rheostat."

Variable transformer - An iron-core transformer with a sliding contact which moves along the exposed turns of the secondary winding to vary the output voltage.

Volt - A practical unit for measuring current pressure in a circuit; the force that will move a current of one ampere through a resistance of one ohm.

Voltage drop - The difference in voltage between two points, caused by the loss of electrical pressure as a current flows through an impedance or resistance. All wire, no matter how low the resistance, shows at least a trace of voltage drop.

Voltage regulator - An electromechanical or electronic device that maintains the output voltage of a device at a predetermined value.

Voltmeter - 1. A test instrument that measures voltage in an electrical circuit. Used to check continuity and determine voltage drop in specific circuits of vehicle electrical systems. 2. An instrument panel gauge that measures system voltage. When the engine's not running, the voltmeter indicates battery voltage, which should be 12 to 13 volts in a 12-volt system. When the engine's running, the voltmeter indicates total system voltage, or the combined voltage output of the alternator and the battery.

VOM (Volt-Ohmmeter) - A two-in-one test instrument. For convenience, a voltmeter and an ohmmeter are mounted in the same case and share a common readout and test leads.

W

Warning light - An instrument panel display used to inform the driver when something undesirable has happened in the monitored circuit or system, such as an overheated engine or a sudden loss of oil pressure.

Watt - The unit for measuring electrical energy or "work." Wattage is the product of amperage multiplied by voltage.

Winding - One or more turns of a wire, forming a coil. Also, the individual coils of a transformer.

Wire - A solid or stranded group of cylindrical conductors together with any associated insulation.

D Understanding wiring diagrams

General information

Wiring diagrams are useful tools when troubleshooting electric circuits. Electrical systems on modern vehicles have become increasingly complex, making a correct diagnosis more difficult. If you take the time to fully understand wiring diagrams, you can take much of the guesswork out of electrical troubleshooting.

If you have an older vehicle, you may find the wiring diagrams difficult to decipher. Frequently, older diagrams give no information on component location, operation or how to read the diagram. Many of them are organized in ways that make tracing wires difficult.

Today's wiring diagrams tend to be more carefully organized, and the cross-references between each portion of the diagram are clearly explained. They often have component locators and some are even in color (to show wire colors).

Reading wiring diagrams

Wiring diagrams either show the entire electrical system on one page, or they split up the electrical system onto multiple pages and have cross-references which tie them all together. When a diagram is split up, there are usually charts with the diagram that explain the cross-referencing system.

Most diagrams show the power source at the top of the page and the grounds at the bottom.

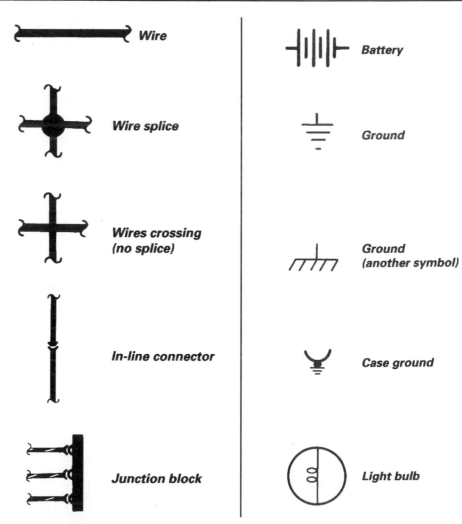

Wire

Wire splice

Wires crossing (no splice)

In-line connector

Junction block

Battery

Ground

Ground (another symbol)

Case ground

Light bulb

Typical wiring diagram symbols – these symbols vary somewhat from manufacturer to manufacturer

Appendix

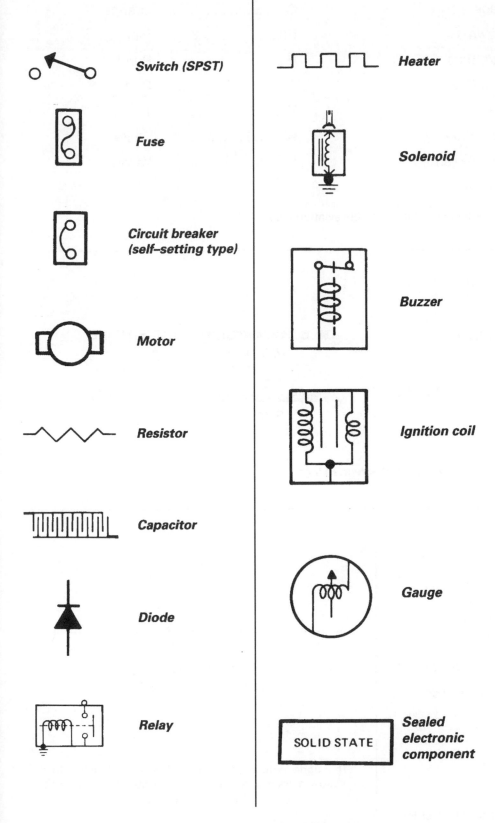

Typical wiring diagram symbols (continued)

Switch (SPST)

Fuse

Circuit breaker (self–setting type)

Motor

Resistor

Capacitor

Diode

Relay

Heater

Solenoid

Buzzer

Ignition coil

Gauge

Sealed electronic component

Components of a wiring diagram

Wiring diagrams can be broken down into three main components: symbols, color codes and wire gage numbers.

Symbols

On wiring diagrams, symbols are used to represent the components of the electrical system. The most obvious symbol is a line to represent a wire. Some other symbols are not so obvious, since they do not necessarily look like the components they're representing. That's because most wiring diagram symbols, which are sometimes called *schematic* symbols, show the way the component functions electrically rather than how it appears physically.

Color codes

Since wiring diagrams are usually in black-and-white, color codes are used to indicate the color of each wire. These codes are normally one or two-character abbreviations. These codes vary somewhat from manufacturer to manufacturer; however, most diagrams include a color code chart so its easy to check the meaning of each code.

Occasionally, manufacturing difficulties cause a manufacturer to deviate slightly from the wiring colors shown on the diagram. If the wire colors at a connector do not match the diagram, and you're sure you're looking at the correct diagram, you can usually identify the incorrect color by comparing all the colors at the connector with the diagram.

Wires are not always solid colors. Often they have markings on them such as a stripe, dots or hash marks. When a diagram shows two colors for a wire, the first color is the basic color of the wire. The second color is the marking.

Wire gage numbers

The wire gage number represents the wire thickness.In a wiring diagram, the gage number for each wire is usually listed either before or after the color code.

Continued on next page

BK	Black	O	Orange
BR	Brown	PK	Pink
DB	Dark Blue	P	Purple
DG	Dark Green	R	Red
GY	Gray	T	Tan
LB	Light Blue	W	White
LG	Light Green	Y	Yellow
N	Natural		

A typical color code abbreviation chart

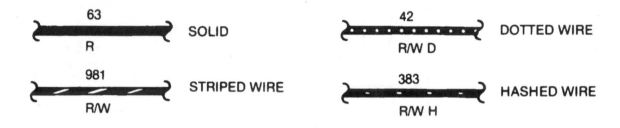

Examples of wire markings

Example: 1.25F - GB

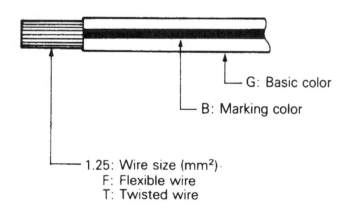

— G: Basic color

— B: Marking color

1.25: Wire size (mm²)
F: Flexible wire
T: Twisted wire

This example code is for a green wire with a black stripe – the 1.25 is the metric wire gage number

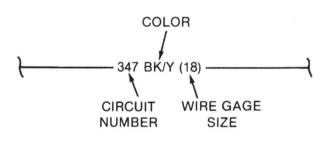

This code indicates the color of the wire (black with a yellow marking), the wire gage number (in AWG) and the circuit number

Notes

Index

Haynes Chevrolet engine overhaul manual

Haynes Automotive Manuals

NOTE: *If you do not see a listing for your vehicle, please visit* **haynes.com** *for the latest product information and check out our* **Online Manuals!**

ACURA
- **12020** Integra '86 thru '89 & **Legend** '86 thru '90
- **12021** Integra '90 thru '93 & **Legend** '91 thru '95
  - Integra '94 thru '00 - *see HONDA Civic (42025)*
  - **MDX** '01 thru '07 - *see HONDA Pilot (42037)*
- **12050** Acura TL all models '99 thru '08

AMC
- **14020** Mid-size models '70 thru '83
- **14025** (Renault) Alliance & Encore '83 thru '87

AUDI
- **15020** 4000 all models '80 thru '87
- **15025** 5000 all models '77 thru '83
- **15026** 5000 all models '84 thru '88
  - **Audi A4** '96 thru '01 - *see VW Passat (96023)*
- **15030** Audi A4 '02 thru '08

AUSTIN-HEALEY
- **Sprite** - *see MG Midget (66015)*

BMW
- **18020** 3/5 Series '82 thru '92
- **18021** 3-Series incl. Z3 models '92 thru '98
- **18022** 3-Series incl. Z4 models '99 thru '05
- **18023** 3-Series '06 thru '14
- **18025** 320i all 4-cylinder models '75 thru '83
- **18050** 1500 thru 2002 except Turbo '59 thru '77

BUICK
- **19010** Buick Century '97 thru '05
  - Century (front-wheel drive) - *see GM (38005)*
- **19020** Buick, Oldsmobile & Pontiac Full-size (Front-wheel drive) '85 thru '05
  - Buick Electra, LeSabre and Park Avenue; Oldsmobile Delta 88 Royale, Ninety Eight and Regency; Pontiac Bonneville
- **19025** Buick, Oldsmobile & Pontiac Full-size (Rear wheel drive) '70 thru '90
  - Buick Estate, Electra, LeSabre, Limited, Oldsmobile Custom Cruiser, Delta 88, Ninety-eight, Pontiac Bonneville, Catalina, Grandville, Parisienne
- **19027** Buick LaCrosse '05 thru '13
  - Enclave - *see GENERAL MOTORS (38001)*
  - Rainier - *see CHEVROLET (24072)*
  - Regal - *see GENERAL MOTORS (38010)*
  - Riviera - *see GENERAL MOTORS (38030, 38031)*
  - Roadmaster - *see CHEVROLET (24046)*
  - Skyhawk - *see GENERAL MOTORS (38015)*
  - Skylark - *see GENERAL MOTORS (38020, 38025)*
  - Somerset - *see GENERAL MOTORS (38025)*

CADILLAC
- **21015** CTS & CTS-V '03 thru '14
- **21030** Cadillac Rear Wheel Drive '70 thru '93
  - Cimarron - *see GENERAL MOTORS (38015)*
  - DeVille - *see GENERAL MOTORS (38031 & 38032)*
  - Eldorado - *see GENERAL MOTORS (38030)*
  - Fleetwood - *see GENERAL MOTORS (38031)*
  - Seville - *see GM (38030, 38031 & 38032)*

CHEVROLET
- **10305** Chevrolet Engine Overhaul Manual
- **24010** Astro & GMC Safari Mini-vans '85 thru '05
- **24013** Aveo '04 thru '11
- **24015** Camaro V8 all models '70 thru '81
- **24016** Camaro all models '82 thru '92
- **24017** Camaro & Firebird '93 thru '02
  - Cavalier - *see GENERAL MOTORS (38016)*
  - Celebrity - *see GENERAL MOTORS (38005)*
- **24018** Camaro '10 thru '15
- **24020** Chevelle, Malibu & El Camino '69 thru '87
  - Cobalt - *see GENERAL MOTORS (38017)*
- **24024** Chevette & Pontiac T1000 '76 thru '87
  - Citation - *see GENERAL MOTORS (38020)*
- **24027** Colorado & GMC Canyon '04 thru '12
- **24032** Corsica & Beretta all models '87 thru '96
- **24040** Corvette all V8 models '68 thru '82
- **24041** Corvette all models '84 thru '96
- **24042** Corvette all models '97 thru '13
- **24044** Cruze '11 thru '19
- **24045** Full-size Sedans Caprice, Impala, Biscayne, Bel Air & Wagons '69 thru '90
- **24046** Impala SS & Caprice and Buick Roadmaster '91 thru '96
  - Impala '00 thru '05 - *see LUMINA (24048)*
- **24047** Impala & Monte Carlo all models '06 thru '11
  - Lumina '90 thru '94 - *see GM (38010)*
- **24048** Lumina & Monte Carlo '95 thru '05
  - Lumina APV - *see GM (38035)*
- **24050** Luv Pick-up all 2WD & 4WD '72 thru '82
- **24051** Malibu '13 thru '19
- **24055** Monte Carlo all models '70 thru '88
  - Monte Carlo '95 thru '01 - *see LUMINA (24048)*
- **24059** Nova all V8 models '69 thru '79

- **24060** Nova and Geo Prizm '85 thru '92
- **24064** Pick-ups '67 thru '87 - Chevrolet & GMC
- **24065** Pick-ups '88 thru '98 - Chevrolet & GMC
- **24066** Pick-ups '99 thru '06 - Chevrolet & GMC
- **24067** Chevrolet Silverado & GMC Sierra '07 thru '14
- **24068** Chevrolet Silverado & GMC Sierra '14 thru '19
- **24070** S-10 & S-15 Pick-ups '82 thru '93, Blazer & Jimmy '83 thru '94,
- **24071** S-10 & Sonoma Pick-ups '94 thru '04, including Blazer, Jimmy & Hombre
- **24072** Chevrolet TrailBlazer, GMC Envoy & Oldsmobile Bravada '02 thru '09
- **24075** Sprint '85 thru '88 & Geo Metro '89 thru '01
- **24080** Vans - Chevrolet & GMC '68 thru '96
- **24081** Chevrolet Express & GMC Savana Full-size Vans '96 thru '19

CHRYSLER
- **10310** Chrysler Engine Overhaul Manual
- **25015** Chrysler Cirrus, Dodge Stratus, Plymouth Breeze '95 thru '00
- **25020** Full-size Front-Wheel Drive '88 thru '93
  - K-Cars - *see DODGE Aries (30008)*
  - Laser - *see DODGE Daytona (30030)*
- **25025** Chrysler LHS, Concorde, New Yorker, Dodge Intrepid, Eagle Vision, '93 thru '97
- **25026** Chrysler LHS, Concorde, 300M, Dodge Intrepid, '98 thru '04
- **25027** Chrysler 300 '05 thru '18, Dodge Charger '06 thru '18, Magnum '05 thru '08 & Challenger '08 thru '18
- **25030** Chrysler & Plymouth Mid-size front wheel drive '82 thru '95
  - Rear-wheel Drive - *see Dodge (30050)*
- **25035** PT Cruiser all models '01 thru '10
- **25040** Chrysler Sebring '95 thru '06, Dodge Stratus '01 thru '06 & Dodge Avenger '95 thru '00
- **25041** Chrysler Sebring '07 thru '10, 200 '11 thru '17 Dodge Avenger '08 thru '14

DATSUN
- **28005** 200SX all models '80 thru '83
- **28012** 240Z, 260Z & 280Z Coupe '70 thru '78
- **28014** 280ZX Coupe & 2+2 '79 thru '83
  - **300ZX** - *see NISSAN (72010)*
- **28018** 510 & PL521 Pick-up '68 thru '73
- **28020** 510 all models '78 thru '81
- **28022** 620 Series Pick-up all models '73 thru '79
  - **720 Series Pick-up** - *see NISSAN (72030)*

DODGE
- **400 & 600** - *see CHRYSLER (25030)*
- **30008** Aries & Plymouth Reliant '81 thru '89
- **30010** Caravan & Plymouth Voyager '84 thru '95
- **30011** Caravan & Plymouth Voyager '96 thru '02
- **30012** Challenger & Plymouth Sapporro '78 thru '83
- **30013** Caravan, Chrysler Voyager & Town & Country '03 thru '07
- **30014** Grand Caravan & Chrysler Town & Country '08 thru '18
- **30016** Colt & Plymouth Champ '78 thru '87
- **30020** Dakota Pick-ups all models '87 thru '96
- **30021** Durango '98 & '99 & Dakota '97 thru '99
- **30022** Durango '00 thru '03 & Dakota '00 thru '04
- **30023** Durango '04 thru '09 & Dakota '05 thru '11
- **30025** Dart, Demon, Plymouth Barracuda, Duster & Valiant 6-cylinder models '67 thru '76
- **30030** Daytona & Chrysler Laser '84 thru '89
  - Intrepid - *see CHRYSLER (25025, 25026)*
- **30034** Neon all models '95 thru '99
- **30035** Omni & Plymouth Horizon '78 thru '90
- **30036** Dodge & Plymouth Neon '00 thru '05
- **30040** Pick-ups full-size models '74 thru '93
- **30042** Pick-ups full-size models '94 thru '08
- **30043** Pick-ups full-size models '09 thru '18
- **30045** Ram 50/D50 Pick-ups & Raider and Plymouth Arrow Pick-ups '79 thru '93
- **30050** Dodge/Plymouth/Chrysler RWD '71 thru '89
- **30055** Shadow & Plymouth Sundance '87 thru '94
- **30060** Spirit & Plymouth Acclaim '89 thru '95
- **30065** Vans - Dodge & Plymouth '71 thru '03

EAGLE
- **Talon** - *see MITSUBISHI (68030, 68031)*
- **Vision** - *see CHRYSLER (25025)*

FIAT
- **34010** 124 Sport Coupe & Spider '68 thru '78
- **34025** X1/9 all models '74 thru '80

FORD
- **10320** Ford Engine Overhaul Manual
- **10355** Ford Automatic Transmission Overhaul
- **11500** Mustang '64-1/2 thru '70 Restoration Guide
- **36004** Aerostar Mini-vans all models '86 thru '97
- **36006** Contour & Mercury Mystique '95 thru '00
- **36008** Courier Pick-up all models '72 thru '82

- **36012** Crown Victoria & Mercury Grand Marquis '88 thru '11
- **36014** Edge '07 thru '19 & Lincoln MKX '07 thru '18
- **36016** Escort & Mercury Lynx all models '81 thru '90
- **36020** Escort & Mercury Tracer '91 thru '02
- **36022** Escape '01 thru '17, Mazda Tribute '01 thru '11, & Mercury Mariner '05 thru '11
- **36024** Explorer & Mazda Navajo '91 thru '01
- **36025** Explorer & Mercury Mountaineer '02 thru '10
- **36026** Explorer '11 thru '17
- **36028** Fairmont & Mercury Zephyr '78 thru '83
- **36030** Festiva & Aspire '88 thru '97
- **36032** Fiesta all models '77 thru '80
- **36034** Focus all models '00 thru '11
- **36035** Focus '12 thru '14
- **36045** Fusion '06 thru '14 & Mercury Milan '06 thru '11
- **36048** Mustang V8 all models '64-1/2 thru '73
- **36049** Mustang II 4-cylinder, V6 & V8 models '74 thru '78
- **36050** Mustang & Mercury Capri '79 thru '93
- **36051** Mustang all models '94 thru '04
- **36052** Mustang '05 thru '14
- **36054** Pick-ups & Bronco '73 thru '79
- **36058** Pick-ups & Bronco '80 thru '96
- **36059** F-150 '97 thru '03, Expedition '97 thru '17, F-250 '97 thru '99, F-150 Heritage '04 & Lincoln Navigator '98 thru '17
- **36060** Super Duty Pick-ups & Excursion '99 thru '10
- **36061** F-150 full-size '04 thru '14
- **36062** Pinto & Mercury Bobcat '75 thru '80
- **36063** F-150 full-size '15 thru '17
- **36064** Super Duty Pick-ups '11 thru '16
- **36066** Probe all models '89 thru '92
  - Probe '93 thru '97 - *see MAZDA 626 (61042)*
- **36070** Ranger & Bronco II gas models '83 thru '92
- **36071** Ranger '93 thru '11 & Mazda Pick-ups '94 thru '09
- **36074** Taurus & Mercury Sable '86 thru '95
- **36075** Taurus & Mercury Sable '96 thru '07
- **36076** Taurus '08 thru '14, Five Hundred '05 thru '07, Mercury Montego '05 thru '07 & Sable '08 thru '09
- **36078** Tempo & Mercury Topaz '84 thru '94
- **36082** Thunderbird & Mercury Cougar '83 thru '88
- **36086** Thunderbird & Mercury Cougar '89 thru '97
- **36090** Vans all V8 Econoline models '69 thru '91
- **36094** Vans full size '92 thru '14
- **36097** Windstar '95 thru '03, Freestar & Mercury Monterey Mini-van '04 thru '07

GENERAL MOTORS
- **10360** GM Automatic Transmission Overhaul
- **38001** GMC Acadia '07 thru '16, Buick Enclave '08 thru '17, Saturn Outlook '07 thru '10 & Chevrolet Traverse '09 thru '17
- **38005** Buick Century, Chevrolet Celebrity, Oldsmobile Cutlass Ciera & Pontiac 6000 all models '82 thru '96
- **38010** Buick Regal '88 thru '04, Chevrolet Lumina '88 thru '04, Oldsmobile Cutlass Supreme '88 thru '97 & Pontiac Grand Prix '88 thru '07
- **38015** Buick Skyhawk, Cadillac Cimarron, Chevrolet Cavalier, Oldsmobile Firenza, Pontiac J-2000 & Sunbird '82 thru '94
- **38016** Chevrolet Cavalier & Pontiac Sunfire '95 thru '05
- **38017** Chevrolet Cobalt '05 thru '10, HHR '06 thru '11, Pontiac G5 '07 thru '09, Pursuit '05 thru '06 & Saturn ION '03 thru '07
- **38020** Buick Skylark, Chevrolet Citation, Oldsmobile Omega, Pontiac Phoenix '80 thru '85
- **38025** Buick Skylark '86 thru '98, Somerset '85 thru '87, Oldsmobile Achieva '92 thru '98, Calais '85 thru '91, & Pontiac Grand Am all models '85 thru '98
- **38026** Chevrolet Malibu '97 thru '03, Classic '04 thru '05, Oldsmobile Alero '99 thru '03, Cutlass '97 thru '00, & Pontiac Grand Am '99 thru '03
- **38027** Chevrolet Malibu '04 thru '12, Pontiac G6 '05 thru '10 & Saturn Aura '07 thru '10
- **38030** Cadillac Eldorado, Seville, Oldsmobile Toronado & Buick Riviera '71 thru '85
- **38031** Cadillac Eldorado, Seville, DeVille, Fleetwood, Oldsmobile Toronado & Buick Riviera '86 thru '93
- **38032** Cadillac DeVille '94 thru '05, Seville '92 thru '04 & Cadillac DTS '06 thru '10
- **38035** Chevrolet Lumina APV, Oldsmobile Silhouette & Pontiac Trans Sport all models '90 thru '96
- **38036** Chevrolet Venture '97 thru '05, Oldsmobile Silhouette '97 thru '04, Pontiac Trans Sport '97 thru '98 & Montana '99 thru '05
- **38040** Chevrolet Equinox '05 thru '17, GMC Terrain '10 thru '17 & Pontiac Torrent '06 thru '09

GEO
- **Metro** - *see CHEVROLET Sprint (24075)*
- **Prizm** - '85 thru '92 see CHEVY (24060), '93 thru '02 see TOYOTA Corolla (92036)
- **40030** Storm all models '90 thru '93
- **Tracker** - *see SUZUKI Samurai (90010)*

(Continued on other side)

Haynes North America, Inc. • (805) 498-6703 • www.haynes.com

Haynes Automotive Manuals (continued)

NOTE: *If you do not see a listing for your vehicle, please visit* **haynes.com** *for the latest product information and check out our* **Online Manuals!**

GMC
- Acadia - *see GENERAL MOTORS (38001)*
- Pick-ups - *see CHEVROLET (24027, 24068)*
- Vans - *see CHEVROLET (24081)*

HONDA
- 42010 Accord CVCC all models '76 thru '83
- 42011 Accord all models '84 thru '89
- 42012 Accord all models '90 thru '93
- 42013 Accord all models '94 thru '97
- 42014 Accord all models '98 thru '02
- 42015 Accord '03 thru '12 & Crosstour '10 thru '14
- 42016 Accord '13 thru '17
- 42020 Civic 1200 all models '73 thru '79
- 42021 Civic 1300 & 1500 CVCC '80 thru '83
- 42022 Civic 1500 CVCC all models '75 thru '79
- 42023 Civic all models '84 thru '91
- 42024 Civic & del Sol '92 thru '95
- 42025 Civic '96 thru '00, CR-V '97 thru '01 & Acura Integra '94 thru '00
- 42026 Civic '01 thru '11 & CR-V '02 thru '11
- 42027 Civic '12 thru '15 & CR-V '12 thru '16
- 42030 Fit '07 thru '13
- 42035 Odyssey all models '99 thru '10
  - Passport - *see ISUZU Rodeo (47017)*
- 42037 Honda Pilot '03 thru '08, Ridgeline '06 thru '14 & Acura MDX '01 thru '07
- 42040 Prelude CVCC all models '79 thru '89

HYUNDAI
- 43010 Elantra all models '96 thru '19
- 43015 Excel & Accent all models '86 thru '13
- 43050 Santa Fe all models '01 thru '12
- 43055 Sonata all models '99 thru '14

INFINITI
- G35 '03 thru '08 - *see NISSAN 350Z (72011)*

ISUZU
- Hombre - *see CHEVROLET S-10 (24071)*
- 47017 Rodeo '91 thru '02, Amigo '89 thru '94 & '98 thru '02 & Honda Passport '95 thru '02
- 47020 Trooper '84 thru '91 & Pick-up '81 thru '93

JAGUAR
- 49010 XJ6 all 6-cylinder models '68 thru '86
- 49011 XJ6 all models '88 thru '94
- 49015 XJ12 & XJS all 12-cylinder models '72 thru '85

JEEP
- 50010 Cherokee, Comanche & Wagoneer Limited all models '84 thru '01
- 50011 Cherokee '14 thru '19
- 50020 CJ all models '49 thru '86
- 50025 Grand Cherokee all models '93 thru '04
- 50026 Grand Cherokee '05 thru '19 & Dodge Durango '11 thru '19
- 50029 Grand Wagoneer & Pick-up '72 thru '91 Grand Wagoneer '84 thru '91, Cherokee & Wagoneer '72 thru '83, Pick-up '72 thru '88
- 50030 Wrangler all models '87 thru '17
- 50035 Liberty '02 thru '12 & Dodge Nitro '07 thru '11
- 50050 Patriot & Compass '07 thru '17

KIA
- 54050 Optima '01 thru '10
- 54060 Sedona '02 thru '14
- 54070 Sephia '94 thru '01, Spectra '00 thru '09, Sportage '05 thru '20
- 54077 Sorento '03 thru '13

LEXUS
- ES 300/330 - *see TOYOTA Camry (92007, 92008)*
- ES 350 - *see TOYOTA Camry (92009)*
- RX 300/330/350 - *see TOYOTA Highlander (92095)*

LINCOLN
- MKX - *see FORD (36014)*
- Navigator - *see FORD Pick-up (36059)*
- 59010 Rear-Wheel Drive Continental '70 thru '87, Mark Series '70 thru '92 & Town Car '81 thru '10

MAZDA
- 61010 GLC (rear-wheel drive) '77 thru '83
- 61011 GLC (front-wheel drive) '81 thru '85
- 61012 Mazda3 '04 thru '11
- 61015 323 & Protegé '90 thru '03
- 61016 MX-5 Miata '90 thru '14
- 61020 MPV all models '89 thru '98
  - Navajo - *see Ford Explorer (36024)*
- 61030 Pick-ups '72 thru '93
  - Pick-ups '94 thru '09 - *see Ford Ranger (36071)*
- 61035 RX-7 all models '79 thru '85
- 61036 RX-7 all models '86 thru '91
- 61040 626 (rear-wheel drive) all models '79 thru '82
- 61041 626 & MX-6 (front-wheel drive) '83 thru '92
- 61042 626 '93 thru '01 & MX-6/Ford Probe '93 thru '02
- 61043 Mazda6 '03 thru '13

MERCEDES-BENZ
- 63012 123 Series Diesel '76 thru '85
- 63015 190 Series 4-cylinder gas models '84 thru '88
- 63020 230/250/280 6-cylinder SOHC models '68 thru '72
- 63025 280 123 Series gas models '77 thru '81
- 63030 350 & 450 all models '71 thru '80
- 63040 C-Class: C230/C240/C280/C320/C350 '01 thru '07

MERCURY
- 64200 Villager & Nissan Quest '93 thru '01
  - *All other titles, see FORD Listing.*

MG
- 66010 MGB Roadster & GT Coupe '62 thru '80
- 66015 MG Midget, Austin Healey Sprite '58 thru '80

MINI
- 67020 Mini '02 thru '13

MITSUBISHI
- 68020 Cordia, Tredia, Galant, Precis & Mirage '83 thru '93
- 68030 Eclipse, Eagle Talon & Plymouth Laser '90 thru '94
- 68031 Eclipse '95 thru '05 & Eagle Talon '95 thru '98
- 68035 Galant '94 thru '12
- 68040 Pick-up '83 thru '96 & Montero '83 thru '93

NISSAN
- 72010 300ZX all models including Turbo '84 thru '89
- 72011 350Z & Infiniti G35 all models '03 thru '08
- 72015 Altima all models '93 thru '06
- 72016 Altima '07 thru '12
- 72020 Maxima all models '85 thru '92
- 72021 Maxima all models '93 thru '08
- 72025 Murano '03 thru '14
- 72030 Pick-ups '80 thru '97 & Pathfinder '87 thru '95
- 72031 Frontier '98 thru '04, Xterra '00 thru '04, & Pathfinder '96 thru '04
- 72032 Frontier & Xterra '05 thru '14
- 72037 Pathfinder '05 thru '14
- 72040 Pulsar all models '83 thru '86
- 72042 Roque all models '08 thru '20
- 72050 Sentra all models '82 thru '94
- 72051 Sentra & 200SX all models '95 thru '06
- 72060 Stanza all models '82 thru '90
- 72070 Titan pick-ups '04 thru '10, Armada '05 thru '10 & Pathfinder Armada '04
- 72080 Versa all models '07 thru '19

OLDSMOBILE
- 73015 Cutlass V6 & V8 gas models '74 thru '88
  - *For other OLDSMOBILE titles, see BUICK, CHEVROLET or GENERAL MOTORS listings.*

PLYMOUTH
- *For PLYMOUTH titles, see DODGE listing.*

PONTIAC
- 79008 Fiero all models '84 thru '88
- 79018 Firebird V8 models except Turbo '70 thru '81
- 79019 Firebird all models '82 thru '92
- 79025 G6 all models '05 thru '09
- 79040 Mid-size Rear-wheel Drive '70 thru '87
  - Vibe '03 thru '10 - *see TOYOTA Corolla (92037)*
  - *For other PONTIAC titles, see BUICK, CHEVROLET or GENERAL MOTORS listings.*

PORSCHE
- 80020 911 Coupe & Targa models '65 thru '89
- 80025 914 all 4-cylinder models '69 thru '76
- 80030 924 all models including Turbo '76 thru '82
- 80035 944 all models including Turbo '83 thru '89

RENAULT
- Alliance & Encore - *see AMC (14025)*

SAAB
- 84010 900 all models including Turbo '79 thru '88

SATURN
- 87010 Saturn all S-series models '91 thru '02
  - Saturn Ion '03 thru '07- *see GM (38017)*
  - Saturn Outlook - *see GM (38001)*
- 87020 Saturn L-series all models '00 thru '04
- 87040 Saturn VUE '02 thru '09

SUBARU
- 89002 1100, 1300, 1400 & 1600 '71 thru '79
- 89003 1600 & 1800 2WD & 4WD '80 thru '94
- 89080 Impreza '02 thru '11, WRX '02 thru '14, & WRX STI '04 thru '14
- 89100 Legacy all models '90 thru '99
- 89101 Legacy & Forester '00 thru '09
- 89102 Legacy '10 thru '16 & Forester '12 thru '16

SUZUKI
- 90010 Samurai/Sidekick & Geo Tracker '86 thru '01

TOYOTA
- 92005 Camry all models '83 thru '91
- 92006 Camry '92 thru '96 & Avalon '95 thru '96
- 92007 Camry, Avalon, Solara, Lexus ES 300 '97 thru '01
- 92008 Camry, Avalon, Lexus ES 300/330 '02 thru '06 & Solara '02 thru '08
- 92009 Camry, Avalon & Lexus ES 350 '07 thru '17
- 92015 Celica Rear-wheel Drive '71 thru '85
- 92020 Celica Front-wheel Drive '86 thru '99
- 92025 Celica Supra all models '79 thru '92
- 92030 Corolla all models '75 thru '79
- 92032 Corolla all rear-wheel drive models '80 thru '87
- 92035 Corolla all front-wheel drive models '84 thru '92
- 92036 Corolla & Geo/Chevrolet Prizm '93 thru '02
- 92037 Corolla '03 thru '19, Matrix '03 thru '14, & Pontiac Vibe '03 thru '10
- 92040 Corolla Tercel all models '80 thru '82
- 92045 Corona all models '74 thru '82
- 92050 Cressida all models '78 thru '82
- 92055 Land Cruiser FJ40, 43, 45, 55 '68 thru '82
- 92056 Land Cruiser FJ60, 62, 80, FZJ80 '80 thru '96
- 92060 Matrix '03 thru '11 & Pontiac Vibe '03 thru '10
- 92065 MR2 all models '85 thru '87
- 92070 Pick-up all models '69 thru '78
- 92075 Pick-up all models '79 thru '95
- 92076 Tacoma '95 thru '04, 4Runner '96 thru '02 & T100 '93 thru '08
- 92077 Tacoma all models '05 thru '18
- 92078 Tundra '00 thru '06 & Sequoia '01 thru '07
- 92079 4Runner all models '03 thru '09
- 92080 Previa all models '91 thru '95
- 92081 Prius all models '01 thru '12
- 92082 RAV4 all models '96 thru '12
- 92085 Tercel all models '87 thru '94
- 92090 Sienna all models '98 thru '10
- 92095 Highlander '01 thru '19 & Lexus RX330/330/350 '99 thru '19
- 92179 Tundra '07 thru '19 & Sequoia '08 thru '19

TRIUMPH
- 94007 Spitfire all models '62 thru '81
- 94010 TR7 all models '75 thru '81

VW
- 96008 Beetle & Karmann Ghia '54 thru '79
- 96009 New Beetle '98 thru '10
- 96016 Rabbit, Jetta, Scirocco & Pick-up gas models '75 thru '92 & Convertible '80 thru '92
- 96017 Golf, GTI & Jetta '93 thru '98, Cabrio '95 thru '02
- 96018 Golf, GTI, Jetta '99 thru '05
- 96019 Jetta, Rabbit, GLI, GTI & Golf '05 thru '11
- 96020 Rabbit, Jetta & Pick-up diesel '77 thru '84
- 96021 Jetta '11 thru '18 & Golf '15 thru '19
- 96023 Passat '98 thru '05 & Audi A4 '96 thru '01
- 96030 Transporter 1600 all models '68 thru '79
- 96035 Transporter 1700, 1800 & 2000 '72 thru '79
- 96040 Type 3 1500 & 1600 all models '63 thru '73
- 96045 Vanagon Air-Cooled all models '80 thru '83

VOLVO
- 97010 120, 130 Series & 1800 Sports '61 thru '73
- 97015 140 Series all models '66 thru '74
- 97020 240 Series all models '76 thru '93
- 97040 740 & 760 Series all models '82 thru '88
- 97050 850 Series all models '93 thru '97

TECHBOOK MANUALS
- 10205 Automotive Computer Codes
- 10206 OBD-II & Electronic Engine Management
- 10210 Automotive Emissions Control Manual
- 10215 Fuel Injection Manual '78 thru '85
- 10225 Holley Carburetor Manual
- 10230 Rochester Carburetor Manual
- 10305 Chevrolet Engine Overhaul Manual
- 10320 Ford Engine Overhaul Manual
- 10330 GM and Ford Diesel Engine Repair Manual
- 10331 Duramax Diesel Engines '01 thru '19
- 10332 Cummins Diesel Engine Performance Manual
- 10333 GM, Ford & Chrysler Engine Performance Manual
- 10334 GM Engine Performance Manual
- 10340 Small Engine Repair Manual, 5 HP & Less
- 10341 Small Engine Repair Manual, 5.5 thru 20 HP
- 10345 Suspension, Steering & Driveline Manual
- 10355 Ford Automatic Transmission Overhaul
- 10360 GM Automatic Transmission Overhaul
- 10405 Automotive Body Repair & Painting
- 10410 Automotive Brake Manual
- 10411 Automotive Anti-lock Brake (ABS) Systems
- 10420 Automotive Electrical Manual
- 10425 Automotive Heating & Air Conditioning
- 10435 Automotive Tools Manual
- 10445 Welding Manual
- 10450 ATV Basics

Over a 100 Haynes motorcycle manuals also available

10/22

Haynes North America, Inc. • (805) 498-6703 • www.haynes.com